Vernon J. Hall

Chemistry and Metallurgy applied to Dentistry

Vernon J. Hall

Chemistry and Metallurgy applied to Dentistry

ISBN/EAN: 9783743419520

Manufactured in Europe, USA, Canada, Australia, Japa

Cover: Foto ©berggeist007 / pixelio.de

Manufactured and distributed by brebook publishing software (www.brebook.com)

Vernon J. Hall

Chemistry and Metallurgy applied to Dentistry

CHEMISTRY AND
METALLURGY
APPLIED TO DENTISTRY

By

Vernon J. Hall, Ph. D.

Professor of Chemistry and Direct=
or of the Chemical Laboratories
in the Dental School and in the
Woman's Medical School of North=
western University.

Published THE TECHNICAL PRESS
by AT EVANSTON, ILLINOIS
MDCCCXCVIII

PREFACE.

This book is not offered to the dental student as an exhaustive treatise on chemistry and metallurgy, but rather as an outline of information which it is hoped will give him a practical knowledge of those facts having an unquestionably important bearing upon dentistry.

In writing this book the author has tried to adhere to three intentions : First, to adapt the course to the time commonly allotted to the study of these subjects; second, to reduce it to a laboratory training, supplemented by the necessary amount of text work; and finally, to eliminate those things which are irrelevant and are not likely to be taught in a practical course. The absence of the smatterings of organic, of physiological and of pharmaceutical chemistry which are so often given in text-books of medical and dental chemistry is a conspicuous feature of this book. The author firmly believes that until the time given to the study of chemistry is increased it is unwise to venture into the field of organic chemistry, and that the time is better spent upon practical problems along inorganic lines, particularly as inorganic chemistry has the wider application in dentistry.

It is assumed, of course, that the student has received instruction either by fully illustrated lectures or by laboratory work, accompanied by lectures in the elementary principles of chemistry, and that he is familiar with ordinary chemical apparatus and with laboratory manipulation. He is then prepared to enter into the course of study outlined in this book.

The author lays little claim to originality except in the general arrangement of the course, in the designing of certain pieces of apparatus, and in the working out of some methods in Part II. He has freely consulted standard books of chemistry and mechanical dentistry, and takes this occasion to acknowledge his indebtedness to the same. He also wishes to thank his assistants, Mr. Wilbur W. Graff and Mr. James E. Remington, for aid in reading proof and in working out details. Finally, he desires to express his sense of obligation to his colleague, Mr. George H. Ellis, and to his former instructor, Professor Abram Van Eps Young, for valuable criticisms and suggestions.　　　　　　　　　　　　　V. J. H.

EVANSTON, ILL., 1899.

CONTENTS.

PART I.
THE METALS. DESCRIPTIVE DETAILS. QUALITATIVE CHEMICAL ANALYSIS.

CHAPTER I.
THE METALS. GENERAL CONSIDERATION.

CHAPTER II.
DESCRIPTIVE DETAILS.

CHAPTER III.
QUALITATIVE CHEMICAL ANALYSIS.

CHAPTER IV.

ANALYSIS OF GROUP I.

CHAPTER V.

ANALYSIS OF GROUP II.

CHAPTER VI.

ANALYSIS OF GROUP III.

CHAPTER VII.

ANALYSIS OF GROUP IV.

CHAPTER VIII.

ANALYSIS OF GROUP V.

CHAPTER IX.

TREATMENT OF METALS AND ALLOYS.

Part II.
CHEMICAL TECHNOLOGY APPLIED TO DENTISTRY.

CHAPTER X.
ALLOYS. GENERAL CONSIDERATION.

CHAPTER XI.
APPARATUS

CHAPTER XII.
REFINING GOLD, SILVER AND MERCURY.

CHAPTER XIII.
DENTAL AMALGAMS AND AMALGAM-ALLOYS.

CHAPTER XIV.

THE ASSAY OF AMALGAM-ALLOYS.

CHAPTER XV.

SOLDERS AND SOLDERING.

CHAPTER XVI.

MISCELLANEOUS ALLOYS.

CHAPTER XVII.

DENTAL CEMENTS.

CHAPTER XVIII.

SPECIAL PROBLEMS.

APPENDIX.

PART I.

The Metals. Descriptive Details. Qualitative Chemical Analysis.

CHAPTER I.

THE METALS. GENERAL CONSIDERATION.

Seventy-four elements are known at the present time. Comparatively few of these substances enter into the composition of common things, while a large number of them are found in such limited quantities that few of their properties are definitely known. It has been customary to divide the elements into two classes, metals and nonmetals. To the former class have been assigned those elements which, with the exception of mercury, are solid at ordinary temperature, have metallic luster, are opaque, are good conductors of heat and electricity and in general possess properties not characteristic of the elements classed as nonmetals, such as hydrogen, oxygen, chlorine, carbon and sulphur. When this distinction was first made but few elements were known, and their classification was based almost entirely upon their physical properties. Researches in chemistry, however, have made it apparent that a sharp line of demarcation cannot be drawn between metals and nonmetals, as some elements possess properties common to both. The proper classification of arsenic, for example, has long been a matter of dispute, owing to the fact that it has a high metallic luster and looks much like a metal and yet is endowed with certain chemical properties which closely ally it to the nonmetallic elements. It is now generally recognized that in this classification of the simple substances the chemical characteristics are of paramount importance; and, as a rule, if an element is base-forming it will be classed as a metal and if acid-forming as a nonmetal, without regard to its physical properties.

The metals are far greater in number, and, from an industrial point of view, of more importance than the nonmetals, as they permit of a more extensive application in the arts. Technically speaking, the more important metals are gold, silver, iron, copper, tin, lead, mercury, platinum, zinc, antimony, nickel, aluminum, bismuth and magnesium. Gold, silver, platinum and mercury are popularly called *noble metals* and the rest are termed *base*. This classification, which had its origin in the alchemistic age, distinguishes those metals having a feeble affinity for oxygen from those which combine with this element with comparative ease at ordinary temperature or on the application of heat. With the exception of the last seven, the metals given in the list above have been known and used from the remotest time.

Occurrence of the Metals.

The question naturally suggests itself, where and in what form are the metals found?

Many metals are very irregularly distributed and,

as already stated, are found in minute quantities only. Fortunately, however, those metals considered of most importance and utility, while constituting but a relatively small portion of the components of the earth's crust, are not diffused in minute quantities, but collected in beds or *veins*. In these veins the metals are found as *minerals*, either in the free state or in combination with other elements in chemical proportions. There are usually several minerals from which a metal can be extracted in profitable quantities, and these are called *ores*. The most abundant ores are the oxides, sulphides and carbonates, examples of which are the two important ores of iron, the oxides, Fe_2O_3 and Fe_3O_4, and the ores of zinc, the sulphide, ZnS, and carbonate, $ZnCO_3$.

Extraction of the Metals From Their Ores.

The various mechanical and chemical methods employed in extracting metals from their ores constitute the art of *metallurgy*. The extraction of metals from oxide ore, as Fe_2O_3, is accomplished by heating at a very high temperature with carbon. As shown in the following equation, the carbon unites with the oxygen, forming carbon dioxide which passes off and metallic iron remains :

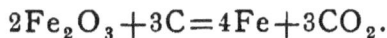

$$2Fe_2O_3 + 3C = 4Fe + 3CO_2.$$

A method not practicable on a large scale but often employed in the laboratory is to pass hydrogen over the heated substance or in some manner to bring hydrogen in close contact with it:

$$CuO + H_2 = Cu + H_2O.$$

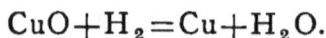

When the ores are sulphides or carbonates they must be converted into oxides before they can be treated as shown above.

Lastly, a method sometimes used when the metal exists in the free state consists in amalgamating it with mercury and subsequently removing the latter by pressure and distillation. This process is more often used in reducing gold and silver ores, and is technically known as the *amalgamation process*.

Physical Properties of the Metals.

As a class the metals are characterized by a metallic luster and by a high degree of opacity except in the case of gold in the form of thin sheets. With the exception of mercury, they are solid at ordinary temperature, but can be melted and in some cases distilled by the application of heat. Many possess a high specific gravity, great hardness, tenacity, ductility and malleability. Some are so ductile that they can be drawn into very fine wire and so malleable that they can be hammered into sheets of great tenuity. Compared with the nonmetallic elements they are good conductors of heat and electricity. The physical properties just cited are not shown equally by all metals; in some instances one or more of them will be found deficient or entirely wanting.

COLOR, LUSTER AND OPACITY.

Most metals are described as white or gray in *color*, but striking exceptions are gold and copper.

Under certain conditions the color and *luster* of metals are greatly modified. This modification usually occurs after oxidation or after tarnishing by certain gases, a new substance being formed on the surface of the metal. Thus steel in being tempered assumes many brilliant colors, and copper when heated slightly exhibits all the colors of the rainbow. The bright surface of lead soon changes to a dull blue in the air and the brilliant luster of silver and many other metals is not long retained in an atmosphere of hydrogen sulphide gas. The alloying of a metal with even small quantities of another often affects the color. An example of this is found in the alloying of gold with silver or copper. Again, metals in a finely divided state are often devoid of luster and possess a color not shown by them when in a dense mass. Precipitated gold is brown and possesses no luster at all; when fused, however, into a mass, it immediately assumes its normal color and luster.

Metals are described as opaque. Gold, however, in very thin sheets transmits a green or purple light.

ODOR AND TASTE.

Although these properties are not very pronounced, a few metals when rubbed or heated give off a peculiar *odor* and under certain conditions are described as having a "metallic" *taste*. Arsenic when heated gives off the odor of garlic.

SPECIFIC GRAVITY.

In *specific gravity*, their weight relative to water, the metals vary greatly. Some are heavy, as gold

and lead, while others are so light that they float
on water, as potassium and sodium.   They vary
from 0.59, the specific gravity of lithium, to 22.0, that
of osmium.   Generally speaking the lighter metals
are the more active chemically.   A variation in
specific gravity from the theoretical mean of the
constituents is often noted in alloys.

FUSIBILITY, VOLATILITY AND CRYSTALLINE FORM.

The most readily *fusible* metal is mercury.   Its
melting point, in reality its freezing point, is about
—39° C.

The alkali metals, potassium and sodium, melt at
the temperature of boiling water; a great number melt
at red heat and a few at white heat only.   Platinum
and some of the rarer metals associated with it in nature require the heat of the oxyhydrogen blowpipe to
effect their fusion.   The melting point of a metal is
often greatly modified by the presence of even traces of
other metals.   Many of the so-called fusible alloys,
composed chiefly of tin, lead and bismuth, melt at
the temperature of boiling water.

Some metals *volatilize* on being heated; arsenic
volatilizes at red heat and can be melted only under
pressure.   Other metals which volatilize readily at
red heat or above are cadmium, zinc, lead and mercury.   The last named is slightly volatile at ordinary
temperature.

Upon being melted and slowly cooled many metals
assume a definite *crystalline form*.   This is particularly
true of those metals melting at comparatively low

temperatures, as bismuth and antimony. Under various other conditions metals show a tendency to crystallize; zinc and arsenic can be crystallized by sublimation; silver has been crystallized from its solutions by galvanic action and at times has been made to assume a crystalline form while in the solid state by repeated heating and cooling and by percussion or other forms of mechanical working. Metals occurring free in nature usually are crystallized.

MALLEABILITY AND DUCTILITY.

Some metals are capable of being rolled, hammered, drawn, or otherwise modified in form by various mechanical means, without becoming disrupted. This property is expressed by the terms *malleability* and *ductility*. Gold is recognized as the most malleable and at the same time the most ductile metal known. It can be rolled or hammered into sheets less than $\frac{1}{300,000}$ of an inch in thickness, and drawn into wire so fine that one mile will weigh less than one gram. Malleability and ductility are by no means proportional in the same metal. Iron, although inferior to tin in malleability, is much more ductile, and can be drawn into very fine wire. In many metals malleability and ductility are wanting altogether. This is exemplified in the cases of bismuth, antimony and other metals more or less crystalline in structure. Both malleability and ductility are influenced by temperature, by mechanical working and by the presence of traces of other metals. As a rule, these properties are increased by an increase of temperature, and if diminished in working the metal, they can be restored by heating and cooling slowly or quickly, usually the former. This is known as *annealing*. The effects of impurities upon a metal are shown in the case of gold. The presence of minute traces of lead, bismuth or antimony in gold greatly impairs its malleability and correspondingly modifies its ductility.

In the following tables the metals are arranged in the order of their malleability and ductility:

TABLE OF MALLEABILITY.		TABLE OF DUCTILITY.	
Most malleable.		*Most ductile.*	
1. Gold.	7. Lead.	1. Gold.	7. Cadmium.
2. Silver.	8. Cadmium.	2. Silver.	8. Aluminum.
3. Aluminum.	9. Zinc.	3. Platinum.	9. Zinc.
4. Tin.	10. Iron.	4. Iron.	10. Tin.
5. Copper	11. Nickel.	5. Nickel.	11. Lead.
6. Platinum.	*Least malleable.*	6. Copper.	*Least ductile.*

HARDNESS, ELASTICITY AND TENACITY.

The metals differ greatly in *hardness*. Potassium and sodium are plastic. Tin and lead are so soft that they can be scratched with the finger nail, while steel can be made so hard that it will scratch glass. The hardening of a metal is usually accomplished by adding to it small quantities of various substances. The presence of one to one and one-half per cent of carbon in steel renders it suitable for making tools, while the addition of even smaller quantities of chromium is said to harden it and improve its quality. Gold and silver in the pure state are too soft for general use; upon alloying them with copper they become quite hard and much more serviceable for coin and for other articles.

In the following table a comparison of the hardness of the principal metals is made with lead, the softest metal in common use.

TABLE OF HARDNESS.

Softest.

Lead	1.0	Antimony	1.8
Tin	1.2	Zinc	1.9
Cadmium	1.4	Platinum	2.0
Aluminum	1.5	Copper	2.4
Bismuth	1.6	Iron	2.4
Gold	1.7	Nickel	2.5
Silver	1.8	*Hardest.*	

In *elasticity*, or the power of recovering original dimensions after being bent, twisted, stretched, etc., the metals are quite deficient. In many cases, however, this property can be induced by alloying, hammering, tempering, etc. When gold is alloyed with a small proportion of platinum it becomes very elastic. When iron is converted into steel and then tempered in a certain manner, it becomes adapted to various uses in which great elasticity is required, as in making sword blades and watch springs.

Closely connected with hardness and elasticity and with malleability and ductility is *tenacity*, the power possessed by metals of resisting forces which tend to separate their particles by tension or crushing. Wide differences are observed in the tenacity of the metals. Lead is the lowest of those recognized as possessing this property, while iron is one of the most tenacious metals known. Some metals show but a slight degree of tenacity and are said to be brittle; for example, antimony, arsenic and zinc. Tenacity is influenced by temperature, by the mechanical working of the metal and by its purity.   As a rule an increase of temperature beyond certain limits reduces the tenacity. Iron and gold heated to 100° C. are somewhat increased in tenacity, but beyond that point a decided decrease is noted.   It is a peculiar fact that, when gold, silver, iron and certain other metals are heated to a red heat and cooled slowly the tenacity of each is greatly diminished.   This undoubtedly is due in part at least to a rearrangement of the molecules.   Silver, for example, ordinarily very tough, can be made to assume a granular form and to become brittle by repeated heating and cooling.   The alloying of metals often reduces the tenacity, but more often increases it; and the mechanical working of a metal can be made to increase or decrease the tenacity in proportion as it induces a fibrous or crystalline texture.   As a rule the more fibrous metals exhibit a high degree of tenacity, while those crystalline in structure are deficient in that respect.   It is obvious that the properties, hardness, elasticity, malleability, ductility and tenacity are more or less intimately connected, and anything affecting one may affect all to a certain extent.

In the following table the metals are arranged in the order of their tenacity, cobalt being the strongest and lead the weakest of the metals:

Most tenacious.

1. Cobalt.	4. Copper.	7. Gold.	10. Cadmium.
2. Nickel.	5. Platinum.	8. Aluminum.	11. Tin.
3. Iron.	6. Silver.	9. Zinc.	12. Lead.

Least tenacious.

CHANGE OF VOLUME WITH TEMPERATURE AND SOLIDIFI-
CATION.

Metals *expand* when heated and, as a rule, a given metal expands uniformly, within certain limits, for equal increments of temperature. As the force exerted in expansion is very great, it is highly important in the industrial arts that the exact amount of expansion which different metals undergo within certain limits of temperature be known and that provision be made for this change of dimension. Thus in building railways, bridges and buildings, the expansion of the steel in warm weather is always taken into consideration. In the table given below the fraction indicates the increase in length, i. e., so-called linear expansion of a rod of the given metal by a rise of temperature from 0° to 100° C. As most metals expand equally in all dimensions, the cubic expansion can be calculated by multiplying the linear expansion by three.

Greatest expansion.

Cadmium ... $\frac{1}{323}$	Copper ... $\frac{1}{593}$
Lead ... $\frac{1}{342}$	Bismuth ... $\frac{1}{617}$
Zinc ... $\frac{1}{313}$	Gold ... $\frac{1}{689}$
Aluminum ... $\frac{1}{433}$	Nickel ... $\frac{1}{737}$
Tin ... $\frac{1}{443}$	Iron (cast) ... $\frac{1}{934}$
Silver ... $\frac{1}{515}$	Antimony ... $\frac{1}{952}$
	Platinum ... $\frac{1}{1123}$

Least expansion.

When metals pass from the liquid to the solid state they suffer a *change of volume*, which in most cases is contraction. Among the common metals which contract are silver, zinc, lead, aluminum, copper and tin. Examples of metals which expand are bismuth and antimony. In employing metals for various purposes in the arts, particularly in making castings of any sort in which perfect detail is required, it is very essential that this tendency to decrease in volume be counteracted. Often this can be accomplished by alloying with other metals, especially with those which expand on solidification. An alloy composed of four parts of lead, one part of tin and one part of antimony furnishes extremely sharp castings, and is employed in the manufacture of type and in the production of dies for swaging purposes.

SPECIFIC HEAT AND CONDUCTING POWER.

The *specific heat* of a substance represents its capacity for heat or, more accurately defined, it represents the quantity of heat required to raise the temperature of a certain weight of a substance a certain number of degrees as compared with that required to raise the temperature of an equal weight of water the same number of degrees. Thus it takes but $\frac{1}{30}$ as much heat to raise the temperature of a pound of mercury ten degrees as it does to raise a pound of water the same number of degrees. Water being taken as the standard, mercury then has a specific heat of $\frac{1}{30}$ or 0.0333.   The capacity of the alkali metals for heat is very great as, in the case of sodium, over five times greater than that of silver and nearly ten times that of gold.

The following table gives the specific heats of the common metals :

Sodium 0.2934	Cadmium..........0.0567
Aluminum.........0.2143	Tin...............0.0562
Potassium.........0.1696	Antimony 0.0508
Iron 0.1123	Mercury 0.0333
Nickel. 0.1086	Gold..............0.0324
Cobalt 0.1070	Lead..............0.0314
Zinc 0.0955	Platinum..........0.0311
Copper............0.0952	Bismuth..........0.0308
Silver.............0.0570	

The metals are the best known *conductors of heat* and *electricity.* In conductivity silver exceeds all other metals and is taken as a standard of comparison. Generally speaking, the best conductors of heat are at the same time the best conductors of electricity. According to Matthiessen, alloying or increasing the temperature of a metal diminishes its power of conducting electricity.

The following tables show the relative conductivity of the more important metals, taking silver as 100 at 0° C.

FOR HEAT.		FOR ELECTRICITY.	
Silver..........100.0		Silver..........100.0	
Copper.........	85.5	Copper.........	97.8
Gold..........	53.2	Gold	76.7
Aluminum......	31.3	Aluminum.......	65.5
Zinc..........	28.1	Zinc...........	29.6
Cadmium	20.1	Cadmium.......	24.4
Tin...........	15.5	Iron...........	14.6
Mercury (liquid)	13.5	Platinum.......	14.5
Iron..........	11.9	Tin............	14.4
Nickel.........	——	Nickel.........	12.9
Lead..........	8.5	Lead..........	8.4
Platinum.......	8.4	Antimony......	3.6
Antimony	4.0	Mercury........	1.8
Bismuth........	1.8	Bismuth........	1.4

MAGNETIC, GALVANIC AND THERMO-ELECTRIC PROPERTIES.

Metals affected by the magnet are said to possess *magnetic quality,* and are divided into two classes, paramagnetic and diamagnetic, according as they are attracted by the magnet or repelled by it.* To the former class belong iron, nickel, cobalt, manganese and chromium. Representatives of the latter class are arsenic, gold, copper, silver, lead, mercury, cadmium, tin, zinc, antimony and bismuth.

Magnetism is communicable; thus a bar of soft iron when brought near a magnet immediately becomes magnetic and retains this property while under the influence of the magnet. Steel acquires magnetism slowly and retains it permanently, forming the so-called permanent magnet. Magnetism is destroyed at red heat, except in the case of cobalt, which, to a slight degree, retains its magnetism until white heat is reached. A natural magnetic substance is the ore of iron known as the magnetic oxide, Fe_3O_4, and commonly called *lodestone.*

When two dissimilar metals in contact with each other are immersed in any conducting liquid which acts upon either of them a current of electricity is produced. Thus when a zinc plate and a platinum plate are placed in sulphuric acid and then connected outside the acid in any manner, as by a copper wire,

---

*A more accurate distinction is: A paramagnetic metal is one which tends to arrange itself parallel to the lines of force about the magnet, and a diamagnetic metal is one which tends to set itself at right angles to these lines.

chemical action takes place, the result being that the zinc is dissolved, that hydrogen is evolved at the platinum plate and that a current of electricity passes through the circuit. This phenomenon is known as *galvanic action*, and the system composed of zinc, platinum and acid arranged as indicated constitutes a galvanic, or more properly a voltaic cell, and embodies the principles upon which common electric batteries are constructed. Of the two metals in the cell, the one attacked by the liquid is called the positive metal or plate and the other the negative metal or plate. In the liquid the current of electricity flows from the positive to the negative metal, while in the connecting wire it flows from the negative to the positive metal.

Galvanic action has an important effect upon certain metals employed for various purposes in the mouth, and will be more fully referred to later.

When two dissimilar metals are connected at two junctions and one of these junctions is heated a current of electricity results. Electricity thus induced by heat is commonly called *thermo-electricity*, and a pair of conductors joined as indicated constitute a thermo-couple. The strength of a thermo-electric current, although in no case very great, is proportional to the difference in temperature between the two junctions, and its direction is dependent upon the metals constituting the thermo-couple. Generally speaking, those metals which possess a decided crystalline structure produce the strongest current; for example, bismuth and antimony.

Chemical Properties of the Metals.

Most metals unite with one another to form *alloys* and with mercury to form *amalgams*. The chemistry of these substances is not very well understood except in cases in which metallic crystals of definite composition are formed. In addition to the compounds just mentioned the metals combine with the nonmetallic elements as follows : With chlorine to form chlorides, with bromine to form bromides, with iodine to form iodides, with sulphur to form sulphides, with oxygen to form oxides, with oxygen and hydrogen to form hydroxides, and with numerous other elements to form compounds of importance.

Metallic oxides and hydroxides, commonly known as *bases*, unite with acids and form a class of compounds, usually neutral, called *salts*. Another chemical characteristic of the metals is their power of replacing hydrogen in acids, the final products again being salts. Examples of common salts are the chlorides, the nitrates, the sulphates, the carbonates, the acetates, etc.

CHAPTER II.

DESCRIPTIVE DETAILS.

LEAD. Symbol, Pb. Combining weight, 206.95. Specific gravity, 11.38. Melting point, 326.2° C. Lead can be easily rolled into sheets, but cannot be drawn into wire. It is very soft, not at all elastic, a poor conductor of heat and electricity, and ranks the lowest in tenacity. It is quite volatile at red heat. When remelted several times it becomes brittle, due to the presence of traces of dissolved oxide. In this condition the metal can be toughened by melting under powdered charcoal. Lead tarnishes in the air, lead monoxide, Pb_2O, being formed. In an atmosphere of hydrogen sulphide gas lead becomes coated with a film of lead sulphide, PbS. It forms four oxides, the most important of which are *litharge*, PbO, and *red lead*, Pb_3O_4. Many lead compounds, particularly *white lead*, $2PbCO_3.Pb(OH)_2$, are used as pigments. Metallic lead is somewhat soluble in impure water, and quite soluble in hot water. It is insoluble in hydrochloric and sulphuric acids, but readily soluble in nitric and acetic acids, lead nitrate, $Pb(NO_3)_2$, and lead acetate, $Pb(C_2H_3O_2)_2$, being formed.

Alloys.* Soft solders, type metal, pewter and fusible alloys. Lead amalgamates readily.

---

*For the composition of the various alloys referred to in this chapter, see chapters on alloys, Part II.

Chief Ore. Galena, PbS. Reduced by roasting until considerable lead oxide and lead sulphate are formed. This mixture is then heated, air being excluded, and the following reactions take place:

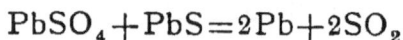

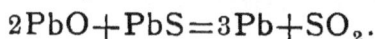

$$PbSO_4 + PbS = 2Pb + 2SO_2$$
$$2PbO + PbS = 3Pb + SO_2.$$

Another method consists in simply heating the ore with iron.

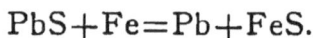

$$PbS + Fe = Pb + FeS.$$

Blowpipe Tests. On charcoal before the blowpipe flame metallic lead gives an easily fusible, soft, gray bead, and an incrustation of oxide, PbO, lemon yellow while hot, sulphur yellow when cold, surrounded by a white border of lead carbonate. Compounds of lead, such as lead chloride or sulphate, are reduced to metallic lead on charcoal by fusing with sodium carbonate.

Confirmatory Reactions. Dissolve the bead of lead in dilute nitric acid.* This solution gives, upon adding hydrochloric acid, a white precipitate of lead chloride, $PbCl_2$, dissolved, after filtering, by pouring hot water over it on the filter paper. Add sulphuric acid to this water solution. A fine white precipitate of lead sulphate, $PbSO_4$, is formed. Filter, and reduce this precipitate to metallic lead by fusing on charcoal with sodium carbonate.

*In dissolving metals, use as little acid as possible and apply heat to promote the reaction. When dissolved, dilute with considerable water or, better still, evaporate nearly to dryness in a porcelain dish and take up the residue with water before applying tests.

SILVER. Symbol, Ag. Combining weight, 107.92. Specific gravity, 10.55. Melting point, 954° C. In degree of malleability and ductility silver is next to gold, and as a conductor of heat and electricity it surpasses all other metals. It is somewhat harder and more tenacious than gold, but still too soft for general use. Silver is hardened by alloying and hammering. It is a pure white metal, not changed by air or water, but readily tarnished by sulphur or its compounds, silver sulphide, Ag_2S, being formed. Silver absorbs about twenty volumes of oxygen when melted and contracts upon cooling, evolving the included oxygen. If the cooling is sudden the oxygen in escaping causes the metal to spirt. This is commonly called the "spitting" or "sprouting" of silver. Silver has few compounds of commercial importance except the nitrate, $AgNO_3$, used in medicine, and the chloride, bromide and iodide, AgCl, AgBr, AgI, all useful in photography. Silver is practically insoluble in hydrochloric acid and in aqua regia, slowly soluble in sulphuric, but readily soluble in nitric acid, silver nitrate, $AgNO_3$, being formed.

Alloys. Coin, jewelry, silver solders and amalgam-alloys, i. e., alloys which when amalgamated are employed in filling cavities in teeth. Silver amalgamates slowly, and the union is attended in most cases with an increase of volume. In certain cases, at least, silver amalgams possess definite chemical composition. Native amalgams of silver frequently are found which are true chemical compounds.

Chief Ores. Argentite, Ag_2S. Silver is often found in paying quantities in galena.

Various methods are employed in treating silver ores. When silver is extracted from galena the *Pattison method* is used. After both lead and silver are reduced to the metallic state, the silver is concentrated by fusing and allowing the lead to crystallize; metallic lead free from silver separates. When an alloy rich in silver is obtained, the remaining lead is separated by the operation of *cupellation*, in which the mixture is heated in bone ash vessels in contact with air; the lead oxidizes and the litharge formed is partly driven off and partly absorbed by the bone ash, leaving the silver in the metallic state.

Another method of extracting silver from its ores is the *amalgamation process*. The ore is roasted with common salt, by which means silver chloride is formed. The mixture is placed in casks and treated with iron and water.

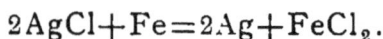

$$2AgCl + Fe = 2Ag + FeCl_2.$$

As observed, the iron reduces the silver chloride to metallic silver. Mercury is next added. This forms with the silver an amalgam, which can be separated from the rest of the mixture. When this amalgam is heated in an iron retort the mercury distills over, leaving the impure silver. Pure silver may be obtained by dissolving the impure varieties in nitric acid and adding common salt.* The resulting silver chloride, after being washed, can be reduced to metallic silver by fusing with sodium carbonate.

$$2AgCl + Na_2CO_3 = 2Ag + CO_2 + O + 2NaCl.$$

*See Chapter XII.

Blowpipe Tests. On charcoal silver fuses easily, giving a white, clear bead, accompanied by no incrustation of oxide. Silver chloride and other compounds of silver are reduced to metallic silver by fusing on charcoal with sodium carbonate.

Confirmatory Reactions. Dissolve the bead in nitric acid and then dilute with water. This solution gives with hydrochloric acid a white, curdy precipitate of silver chloride, AgCl, insoluble in hot water (distinction from lead chloride) but soluble on the filter paper, in hot ammonium hydroxide. If nitric acid is added to this ammoniacal solution the precipitate of silver chloride reappears. Filter and reduce to metallic silver by fusing on charcoal with sodium carbonate.

MERCURY. Symbol, Hg. Combining weight, 200. Specific gravity, 13.59. Melting point (freezing), —38.8° C. Boiling point, 357.2° C. Mercury, also known as *quicksilver*, is the only metal liquid at ordinary temperature. When pure it has a brilliant silver color, but a slight trace of impurity is shown by the formation of a scum on its surface. It does not oxidize, but slowly volatilizes in the air at ordinary temperature. One of the peculiarities of mercury is its tendency to unite with other metals and make the class of substances known as *amalgams*. Mercury forms two oxides, mercurous oxide, Hg_2O, and mercuric oxide, HgO. The latter is commonly called *red precipitate*. In addition to these compounds mercury forms others of importance, particularly the chlorides,

Hg_2Cl_2, commonly called *calomel*, and $HgCl_2$, known as *corrosive sublimate*, a powerful disinfectant. Mercuric sulphide, HgS, often called *vermilion* or *cinnabar*, is employed in coloring rubber. When used in coloring vulcanizable rubber for artificial dentures, vermilion free from metallic mercury, red lead, etc., with which it is sometimes contaminated, should be employed. Mercury is insoluble in hydrochloric and sulphuric acids but soluble in aqua regia, forming mercuric chloride, $HgCl_2$, and in nitric acid, forming mercurous and mercuric nitrates, $HgNO_3$ and $Hg(NO_3)_2$.

Alloys. Under certain conditions mercury unites with nearly all metals to form *amalgams*, some of which, however, are very unstable substances. Mercury when mixed with certain alloys, composed chiefly of silver and tin, forms an amalgam widely used in filling cavities in teeth.

Chief Ore. Cinnabar, HgS. Reduced by roasting in the air and then refined by distilling with lime:

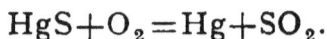

$$HgS + O_2 = Hg + SO_2.$$

Blowpipe Tests. Metallic mercury quickly volatilizes on charcoal and gives a metallic mirror when heated in a glass tube sealed at one end.

Confirmatory Reactions.

Mercurous Compounds. Dissolve some mercury in very dilute nitric acid. Hydrochloric acid added to this solution produces a white precipitate of mercurous chloride, Hg_2Cl_2, insoluble in hot water (distinction from lead chloride). Filter and treat on the filter paper with ammonium hydroxide. The precipitate blackens, but does not dissolve (distinction from silver chloride), a complex compound, mercurous ammonium chloride, NH_2Hg_2Cl, being formed.

Mercuric Compounds. Dissolve some mercuric nitrate or chloride in hot water. Add hydrochloric acid. No precipitate is formed. Add hydrogen sulphide gas. In time a black precipitate of mercuric sulphide, HgS, is obtained. Filter, dry and heat with sodium carbonate in a glass tube sealed at one end. A deposit of metallic mercury will appear on the cool part of the tube.

ARSENIC. Symbol, As. Combining weight, 75. Specific gravity, 4.71. Melting point (under pressure) between silver and antimony. Arsenic is usually classed as a nonmetal, although in some respects it resembles the metals. It is sometimes black and sometimes steel-gray in color. It is brittle and pulverizable. It slowly oxidizes in moist air and when heated it volatilizes without melting. At a high temperature it burns with a bluish flame, emitting the odor of garlic, and becoming arsenious oxide, As_2O_3. Metallic arsenic is not poisonous, but the "white" arsenic, or arsenious oxide, is extremely so. Metallic arsenic is quite uncommon as a metal and is of little value. Arsenious oxide has many uses, particularly in medicine and dentistry. In the latter it is used as a devitalizing agent. Metallic arsenic is but slightly acted upon by common acids in the cold. It dissolves in hot concentrated nitric acid and in aqua regia, forming arsenic acid, H_3AsO_4. Arsenious oxide is nearly insoluble in cold water, but soluble in hydrochloric and sulphuric acids, in alkalies and in the fluids of the stomach.

Alloys. Arsenic forms no alloys of importance. It is present in shot to the extent of one-half per cent, and is a very persistent impurity in zinc and tin.

Chief Ore. Arsenical pyrites, FeAsS. Reduced by heating, air being excluded.

$$FeAsS = As + FeS.$$

Blowpipe Tests. Metallic arsenic volatilizes without fusing, giving the odor of garlic, and a white incrustation of arsenious oxide, As_2O_3, on the charcoal.

Compounds, as arsenious oxide, when heated with charcoal in a glass tube sealed at one end, give a deposit of gray metallic arsenic on the cool part of the tube.

Confirmatory Reactions. Dissolve some arsenious oxide in hydrochloric acid with the aid of heat. Dilute with water and add hydrogen sulphide gas. A yellow precipitate of arsenious sulphide, As_2S_3, is formed. Filter. Place the precipitate and paper in a porcelain dish and digest with yellow ammonium sulphide. The precipitate dissolves.

Marsh's Test for Arsenic. A test for arsenic more delicate than those given above is that known as Marsh's test. In making this test a modified hydrogen generator (Fig. 25) commonly known as Marsh's apparatus is employed. The flask is charged with zinc and hydrochloric acid as usual in generating hydrogen. After the evolution of gas has continued until the air is expelled, a towel is wrapped about the apparatus as a precautionary measure, and the hydrogen is lighted at the small opening of the exit tube. The solution to be tested is poured into the generator through the funnel tube, and a piece of cold porcelain is held in contact with the flame. The compound, arsenious hydride, formed in the generator passes out and burns at the opening, depositing metallic arsenic on the porcelain. Arsenic spots are of a steel gray luster and soluble in sodium hypochlorite.

ANTIMONY. Symbol, Sb. Combining weight, 120. Specific gravity, 6.7. Melting point, 432° C. Antimony is a very brittle and readily pulverizable metal. It is bluish white in color, possesses a crystalline structure and in general appearance resembles bismuth. It does not readily change in dry air at ordinary temperature, but when heated to a red heat it yields antimonious oxide, Sb_2O_3. Hydrogen sulphide only slightly tarnishes it. There are no compounds of antimony of much importance. The metal, however, is widely used in alloys. In its chemical properties antimony resembles arsenic. It is practically insoluble in hydrochloric and sulphuric acids except when the latter is concentrated and boiling. It is oxidized by nitric acid forming antimonic acid, H_3SbO_4. Aqua regia dissolves it, forming antimony chloride, $SbCl_3$. It is soluble also in tartaric acid but not in alkalies.

Alloys. Britannia metal, type metal, Babbitt metal. Antimony hardens alloys and causes them to expand on cooling, thus filling the mold. It forms an amalgam which decomposes in air and water.

Chief Ore. Stibnite, Sb_2S_3. Reduced by first roasting and then heating with carbon.

$$Sb_2O_4 + 4C = 2Sb + 4CO.$$

Blowpipe Tests. Antimony fuses easily and covers the charcoal with a white incrustation of antimonious oxide, Sb_2O_3. If the bead is dropped from the charcoal onto the table it skips about, leaving a white trail.

Confirmatory Reactions. Dissolve the bead in a little aqua regia, dilute with water, disregard any precipitate, and add hydrogen sulphide gas. An orange red precipitate of antimonious sulphide, Sb_2S_3, is formed. Dissolve, like arsenious sulphide, by boiling with yellow ammonium sulphide.

Marsh's Test for Antimony. The Marsh test for antimony is conducted in the same manner as the corresponding test for arsenic. Hydrogen is generated in the apparatus and after the air is expelled the solution to be tested is poured in; a towel is wrapped about the apparatus, the gas is ignited and a piece of cold porcelain is held against the flame. Antimonious hydride is decomposed and yields a brown or black velvety spot of metallic antimony, insoluble in sodium hypochlorite.

TIN. Symbol, Sn. Combining weight, 119. Specific gravity, 7.20. Melting point, 232.7° C. Tin is a very white metal, much resembling silver in this respect, but not so bright in color. It is soft and malleable, capable of being hammered into very thin foil, but not very ductile or tenacious. It does not readily change in the air at ordinary temperature, but it quickly tarnishes in hydrogen sulphide gas, stannous sulphide, SnS, being formed. It burns at a white heat, stannic oxide, SnO_2, resulting. When crushed or bent tin emits a peculiar crackling sound. Few compounds of tin are of importance. *Mosaic gold*, used as a bronze powder, is tin sulphide, SnS_2. Varieties of pure tin are *Banca tin* and *block tin*. Tin dissolves rapidly in pure concentrated hydrochloric and sulphuric acids, forming stannous chloride, $SnCl_2$, and stannous sulphate, $SnSO_4$. Nitric acid rapidly converts it into a white, insoluble substance, metastannic acid, $H_{10}Sn_5O_{15}$ (variable). Aqua regia dissolves tin as stannic chloride, $SnCl_4$.

Alloys. Amalgam-alloys, bronze, soft solders, type metal, Britannia metal and fusible alloys. Tin amalgamates very readily, the product being used in "silvering" mirrors. The combination of tin and mercury is attended by a decrease in volume.

Chief Ore. Tin stone, SnO_2. The ore is roasted to remove arsenic and then reduced by heating with coal.

$$SnO_2 + 2C = Sn + 2CO.$$

Blowpipe Test. Tin fuses easily, giving an incrustation of stannic oxide, SnO_2, yellow while hot, white when cold. Tin is very oxidizable.

Confirmatory Reactions. Dissolve the bead in hydrochloric acid, dilute with water and to a portion add hydrogen sulphide gas. A brown precipitate of stannous sulphide, SnS, appears, soluble, like the sulphides of arsenic and antimony, in hot yellow ammonium sulphide. To another portion add a solution of mercuric chloride; a white precipitate of mercurous chloride, Hg_2Cl_2, appears, and this is taken as evidence of tin. The reaction involved is as follows: When mercuric chloride, $HgCl_2$, a soluble substance, is added to stannous chloride, $SnCl_2$, a reducing agent, the latter becomes oxidized to stannic chloride, $SnCl_4$, also soluble, and the former is reduced to an insoluble form, mercurous chloride, Hg_2Cl_2, hence it precipitates and proves the presence of tin in the solution.

BISMUTH. Symbol, Bi. Combining weight, 208.9. Specific gravity, 9.76. Melting point, 268.3° C. Bismuth is a very brittle metal, slightly reddish yellow in color. It possesses a crystalline structure and resembles antimony except in color. Under ordinary conditions it remains unchanged in the air but at a red heat it burns with a blue flame, the oxide of bismuth, Bi_2O_3, being formed. In the presence of moisture, hydrogen sulphide forms bismuth sulphide, Bi_2S_3. There are few important compounds of bismuth except bismuth subnitrate, $Bi(OH)_2NO_3$, used in medicine. Bismuth is insoluble in hydrochloric acid and but slightly soluble in sulphuric. It is very soluble in nitric acid, forming bismuth nitrate, $Bi(NO_3)_3$.

Alloys. Bismuth is chiefly used in the so-called fusible alloys, to which it imparts a low melting point. Examples of fusible alloys are Wood's, Newton's and Rose's metals. Bismuth amalgamates readily and is sometimes used to adulterate mercury. One part of bismuth in eight thousand parts of mercury can be detected and separated as a black powder by shaking the mercury in a test tube or other suitable vessel.

Chief Ores. Bismuth occurs native, also as bismuth glance, Bi_2S_3, and as bismuth ocher, Bi_2O_3. The metal is reduced from the oxide by fusing with carbon. Native metallic bismuth is separated from the ore by heating in inclined iron retorts. The metal melts at a low heat, and runs from a small opening in the lower end of the retort.

Blowpipe Test. On charcoal bismuth fuses readily, giving a brittle bead, which distinguishes it from lead, and an incrustation of bismuth oxide, Bi_2O_3, orange yellow while hot, lemon yellow when cold.

Confirmatory Reactions. Dissolve the bead in nitric acid and dilute with water. To a portion add ammonium hydroxide to alkaline reaction. A white precipitate of bismuth hydroxide, $Bi(OH)_3$, is formed. Filter, and pour over the precipitate on the filter

paper hot *stannite*.* The precipitate turns black. To
the second portion add hydrogen sulphide gas. A
black precipitate of bismuth sulphide, Bi_2S_3, is
formed, insoluble in ammonium sulphide (distinction
from arsenic, antimony and tin). Filter and digest
with nitric acid in a porcelain dish. The precipitate
dissolves.

COPPER. Symbol, Cu. Combining weight, 63.6.
Specific gravity, 8.95. Melting point, 1054° C. Specific gravity and melting point are both variable. Copper is a red metal of brilliant luster, not readily tarnished except in moist air and in hydrogen sulphide gas.
In the former case basic carbonate, $CuCO_3.CuO_2H_2$,
is formed and in the latter copper sulphide, CuS.
Copper is quite a difficult metal to fuse. When
heated a brilliant film of oxide covers its surface.
Copper is one of the most malleable metals, but
only approaches iron in ductility and tenacity. As a
conductor of heat and electricity it is next to silver.
Important compounds of copper are cuprous oxide,
Cu_2O, cupric oxide, CuO, copper sulphate, $CuSO_4$,
and copper arsenite, commonly known as *Scheele's
green*, a very poisonous substance used as a pigment.
Copper dissolves in nitric acid forming copper
nitrate, $Cu(NO_3)_2$, and in sulphuric, forming copper
sulphate, $CuSO_4$.

---

*To make *stannite* add potassium hydroxide to a drop of
stannous chloride, in a test tube, until the precipitate at first
formed redissolves, and the liquid is strongly alkaline.

Alloys. Coin, brass, bronze, gun metal, bell metal, aluminum bronze, German silver, Babbitt metal, amalgam-alloys, and in various alloys made to imitate gold. It amalgamates if its surface is perfectly clean. The precipitated copper, obtained by the action of metallic iron or zinc in a weak solution of copper sulphate, is used in dentistry as a filling material when amalgamated.

Chief Ores. Copper is found native, also as copper glance, Cu_2S, and as malachite, $CuCO_3.CuO_2H_2$. Oxide and carbonate ores are reduced by smelting with coal. The reduction of sulphide ores is complicated owing to the difficulty with which they are oxidized. The ore is first roasted in the air then with a siliceous flux and carbon, copper *matte* being formed. This process is repeated until the ore is nearly pure copper sulphide and then it is reduced to metallic copper by heating in the air, the oxide formed reacting as follows :

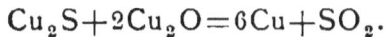

$$Cu_2S + 2Cu_2O = 6Cu + SO_2.$$

Another important method of reducing copper is by electrolysis.

Blowpipe Tests. Copper fuses with difficulty on charcoal before the ordinary blowpipe flame. In the oxidizing flame it gives a red bead. Copper and its compounds color the flame green.

Confirmatory Reactions. Dissolve the bead in nitric acid. A blue solution results. Dilute and pass in hydrogen sulphide gas. This gives a black precipitate of copper sulphide, CuS, not readily soluble in ammonium sulphide (distinction from arsenic, antimony and tin). Filter and digest with nitric acid. Again filter, and to the filtrate add ammonium hydroxide to alkaline reaction. This gives a deep blue solution readily decolorized by potassium cyanide, a double cyanide of copper being formed. If hydrogen sulphide is now added to the solution no precipitate of sulphide is obtained (method of separating copper and cadmium).

CADMIUM. Symbol, Cd. Combining weight, 112. Specific gravity, 8.66. Melting point, 320° C. Cadmium is a white, crystalline metal, somewhat softer than aluminum and harder than tin. It resembles tin in its general appearance, and crackles like tin when crushed. It is quite malleable and ductile but not very tenacious. Cadmium does not readily tarnish except in an atmosphere of hydrogen sulphide gas, due to the formation of cadmium sulphide, CdS. When heated to a low red heat it burns, forming cadmium oxide, CdO. It is somewhat uncommon in the metallic condition and forms few compounds of importance. Cadmium is soluble in hot dilute hydrochloric or sulphuric acid, but more soluble in nitric acid, cadmium nitrate, $Cd(NO_3)_2$, being formed.

Alloys. Fusible alloys and amalgam-alloys. As a constituent of the latter cadmium has fallen into disfavor, owing in part to its tendency to disintegrate and tarnish in the mouth. Cadmium amalgamates readily.

Chief Ore. Greenockite, CdS, found more or less associated with zinc. Reduced by heating with carbon and refined by distillation.

Blowpipe Tests. On charcoal cadmium burns without melting, giving an incrustation of reddish brown cadmium oxide, CdO.

Confirmatory Reactions. Dissolve some of the metal in nitric acid, dilute with water and add hydrogen sulphide gas. A yellow precipitate of cadmium sulphide, CdS, is obtained, resembling in color the corresponding precipitate of arsenic but distinguished from the latter by its insolubility in ammonium sulphide. The addition of potassium cyanide, previous to hydrogen sulphide gas, does not interfere with the precipitation of cadmium as cadmium sulphide (distinction from copper).

ALUMINUM. Symbol, Al. Combining weight, 27. Specific gravity, 2.58. Melting point, 700° C. Aluminum resembles tin in color and silver in hardness and tenacity. When pure it does not tarnish in the air nor in the presence of hydrogen sulphide gas. At a high temperature it oxidizes, forming aluminum oxide, Al_2O_3. Aluminum is one of the lightest metals, and at the same time it rivals steel in strength. It is malleable and ductile, but becomes brittle upon working, and must be frequently annealed. The best temperature for working is between 100° and 150° C. The conducting power of aluminum for heat is about one-third and for electricity about two-thirds that of silver. In addition to the natural products referred to later, aluminum forms few compounds of importance. *Common alum* has the composition $K_2SO_4 \cdot Al_2(SO_4)_3 + 24H_2O$. Aluminum is insoluble in cold nitric and slowly soluble in cold sulphuric acid. Hot, concentrated nitric acid dissolves it slowly. It dissolves readily in hydrochloric acid, forming aluminum chloride, Al_2Cl_6, and in alkali solutions, as potassium hydroxide, in this case forming an aluminate, $K_2Al_2O_4$. In the presence of sodium chloride, organic acids attack aluminum somewhat.

Alloys. Aluminum bronze, aluminum steel, Hercules metal. Aluminum and silver form an important alloy used in making physical and chemical instruments. Pure aluminum as well as the alloy of aluminum and copper, i. e., aluminum bronze, is often used as the base for artificial dentures. Aluminum unites with mercury under certain conditions. The product is not stable, but decomposes into aluminum oxide and mercury, evolving heat. When moisture is present an aluminum amalgam greatly increases in volume and evolves hydrogen, due to the decomposition of the water. This same phenomenon is observed when alloys containing as little as one to two per cent of aluminum are amalgamated. Owing to the facts just stated aluminum is not a desirable constituent of amalgam-alloys, although it is sometimes added.

Chief Ores. Bauxite, $Al_2O_3 + 3H_2O$, cryolite, Na_3AlF_6, orthoclase, $KAlSi_3O_8$. The reduction is very complicated, and not yet entirely satisfactory. Probably the best strictly chemical method consists in heating the double chloride of aluminum and sodium with metallic sodium.

$$AlCl_3 . NaCl + 3Na = Al + 4NaCl.$$

Electrolytic methods have displaced this for commercial purposes.

Blowpipe Tests. On charcoal aluminum melts quite readily, but does not flow into a bead. If heated on asbestos in the oxidizing flame and then moistened with cobalt nitrate, it appears blue upon cooling.

Confirmatory Reactions. Dissolve the metal in hydrochloric acid. To a portion of the solution add potassium hydroxide. A white, gelatinous precipitate of aluminum hydroxide, $Al_2(OH)_6$, is formed, soluble in an excess of the reagent, forming potassium aluminate, $K_2Al_2O_4$. To another portion add ammonium hydroxide; the same precipitate is formed, but in this case it is not soluble in an excess of the reagent.

Aluminum Silicates. Among the important minerals containing aluminum silicate, and occurring with the exception of the last named in enormous quantities in nature are *felspar*, *kaolin*, *clay* and *ultramarine.* The first three substances enter into the composition of porcelain, brick, earthenware, etc., while the last is an important coloring matter.

Felspars. The more common felspars are orthoclase, $KAlSi_3O_8$, and albite, $NaAlSi_3O_8$. By the action of natural agencies these substances are decomposed and large beds of more or less pure clay result.

Kaolin or China Clay. This is the purest form of

aluminum silicate found in nature. It has the approximate composition $Al_4(SiO_4)_3 + 4H_2O$.

Clays. Ordinary clays are impure varieties of aluminum silicate. They are often highly colored with compounds of iron. Clay mixed with considerable calcium carbonate is commonly called *marl.* Other clays of indefinite composition are *loam, yellow ocher, Fuller's earth,* etc.

Ultramarine. This is a substance highly valued as a coloring matter. It consists of a silicate of sodium and aluminum together with certain compounds of sulphur. As already stated, it is found in limited quantities in nature and was formerly very expensive, but is now produced artificially in great quantities. In addition to the blue, a green, a red and a yellow can be produced.

Porcelain. In the manufacture of porcelain, kaolin is used. When mixed with water it can be made into a plastic mass, capable of being molded into any required form. Articles made from clay alone, however, are liable to crack on drying, due to included water. This difficulty is overcome by adding some variety of silica (silex), which renders the mass more open and more easily desiccated. To compensate for the loss of tenacity occasioned by adding silica, a certain quantity of felspar is added. This vitrifies and acts as a cement. The ingredients are thoroughly mixed in the required proportions, molded, dried, and then subjected to a red heat. This gives a porous, unglazed product. The article is next covered with the glazing material and burned again, at a white heat.

This produces the glaze. Glazing materials usually consist of the same materials as compose the body of the porcelain, mixed with a large proportion of fel spar to render them very fusible. Porcelain is colored by means of various metallic oxides. In the making of dental porcelain a relatively high proportion of the felspar is used and the desired shades of enamel are obtained by the use of finely divided platinum, titanium oxide, purple of Cassius, etc.

ZINC. Symbol, Zn. Combining weight, 65.3. Specific gravity, 7.14. Melting point, 450° C. Zinc is a bluish white metal, not readily tarnished in air or water. When heated to a low red heat it burns with a bluish flame, forming zinc oxide, ZnO. Ordinarily zinc is lacking in malleability and ductility. Between 100° and 150° C., however, it can be rolled into sheets and drawn into wire. At 200° C. it becomes so brittle that it can be powdered. In dentistry zinc is widely employed in making dies upon which metal plates are swaged. Its low melting point, its hardness and its toughness render it particularly adaptable to this purpose, although its tendency to contract upon cooling is objectionable. After being melted several times the metal often becomes very brittle owing to the presence of dissolved zinc oxide or to contamination with iron from the ladle in which it is melted. In this condition zinc may be purified by throwing some dry ammonium chloride upon its surface when in the molten state. The tenacity of zinc is twice that of lead. Its most important compound is zinc oxide, widely used as a pigment and in dental cement powders, etc. Zinc dissolves in all common acids, forming the following well-known substances: Zinc chloride, $ZnCl_2$, zinc sulphate, $ZnSO_4$, and zinc nitrate, $Zn(NO_3)_2$. The impure varieties of zinc and those alloyed with copper or platinum are acted upon by acids more readily than the pure metal. Like aluminum, zinc dissolves in alkali solutions; for example, in potassium hydroxide, forming a zincate, K_2ZnO_2.

Alloys. Zinc is a constituent of brass, various solders, amalgam-alloys, etc. It amalgamates, the product being used in preserving the zinc plates in batteries. Zinc is also widely used in coating iron, forming the well-known and useful *galvanized iron.*

Chief Ore. Zinc blende, ZnS. Reduced by roasting and then distilling with carbon. Zinc is usually associated with arsenic in nature, so that commercial zinc and zinc oxide will seldom be found free from this impurity.

Blowpipe Tests. Zinc does not melt on charcoal but burns, giving an incrustation of zinc oxide, ZnO, yellow while hot, white when cold. If moistened with cobalt nitrate the mass becomes green.

Confirmatory Reactions. Dissolve some zinc in hydrochloric acid and add potassium hydroxide. A white precipitate of zinc hydroxide, $Zn(OH)_2$, forms and readily redissolves like aluminum hydroxide in an excess of the reagent, potassium zincate, K_2ZnO_2, being formed. If then ammonium sulphide is added, a white precipitate of zinc sulphide, ZnS, is obtained.

IRON. Symbol, Fe. Combining weight, 56. Specific gravity, 7.79-7.84. Melting point of gray cast iron, 1275° C., and of cast steel, 1375° C. The physical properties of iron vary with the method of manufacture. Chemically pure iron is soft, lustrous and silver gray in color. Air and moisture are very destructive to iron, producing hydrated ferric oxide, $Fe_2O_3.3H_2O$ (variable), commonly known as *rust*. To preserve iron it is painted, tinned, galvanized, etc. When strongly heated in the air iron becomes coated with a layer of magnetic oxide, Fe_3O_4. In tenacity iron exceeds all other metals but nickel and cobalt. It is one of the most ductile metals, but somewhat deficient in malleability, ranking with zinc. When heated, however, its ductility and malleability are greatly increased. Iron is very infusible except at the highest heat of the blast furnace. It possesses the peculiar property of softening at white heat, and in this condition can be welded. Pure iron is attracted by the magnet, but does not retain magnetism. Permanent magnets are made of steel. The common oxides of iron are ferrous oxide, FeO, and ferric oxide, Fe_2O_3. Well-known compounds of iron are ferric chloride, Fe_2Cl_6, and ferrous sulphate, $FeSO_4$, both used in medicine. Iron dissolves in all common acids, forming with hydrochloric, ferrous chloride, $FeCl_2$, and with sulphuric, ferrous sulphate, $FeSO_4$. With nitric acid iron forms ferric nitrate, $Fe_2(NO_3)_6$.

Alloys. Iron forms few alloys of importance except its combination with nickel, manganese, chromium, etc., in steels. Tinned and galvanized iron have already been referred to. An iron amalgam can be obtained by adding sodium amalgam to a solution of an iron salt.

Chief Ores. Hematite, Fe_2O_3, magnetite, Fe_3O_4, and clay ironstone, $FeCO_3$. Iron is reduced by heating in special blast furnaces. The ore, with alternating layers of coal and limestone or other flux, is placed in the top of a tall cylindrical furnace. An intense heat is applied and air is blown in through pipes known as tuyères. Clay, sand and other impurities unite with the limestone and form a fusible slag. The iron is reduced, and uniting with carbon and silicon forms a fusible mass, which settles in the furnace and later is molded in sand troughs into pigs. Iron thus obtained is known as *pig iron* or *cast iron*, and contains carbon, sulphur, silicon, etc.

Wrought iron is made by subjecting pig iron to the *puddling process*, which consists in melting the iron in contact with air, and thus oxidizing the carbon, silicon and other constituents. Wrought iron is very malleable, and increases in this respect as the percentage of carbon decreases.

Steel. The most important and rapid process for making steel is that known as the *Bessemer process*. It consists in burning out the carbon and silicon in cast iron by blowing air through the molten metal, and then adding pure cast iron in such a quantity as to introduce the required percentage of carbon. The steel is next cast into ingots. For many purposes the addition of small quantities of phosphorus, manga-

nese, chromium, tungsten, etc., improves its quality.
Steel is grayish white in color and is capable of
taking a high polish. When heated and cooled suddenly it becomes extremely hard and brittle. By
properly reheating hardened steel to certain temperatures below red heat and then cooling suddenly, it is
possible to produce steel of any required hardness.
This is known as *tempering*.

Blowpipe Tests. On charcoal iron is infusible. Iron
compounds reduced on charcoal with sodium carbonate give the magnetic oxide, Fe_3O_4, which is attracted
by the magnet.

Confirmatory Reactions. Dissolve some iron in
hydrochloric acid. Add chlorine water and boil.* To
a portion add ammonium hydroxide to alkaline reaction. A reddish precipitate of ferric hydroxide,
$Fe_2(OH)_6$, is formed. Filter, dissolve in hydrochloric
acid and add potassium sulphocyanate. A red solution is formed. This is a very delicate test. To
another portion add potassium hydroxide; ferric
hydroxide is again formed, but unlike the corresponding precipitates of zinc and aluminum, it is insoluble
in an excess of the reagent.

MANGANESE. Symbol, Mn. Combining weight,
55. Specific gravity, 7.14-7.20. Manganese as a
metal is not very common, and is used chiefly to improve the quality of steel. It is hard, brittle and resembles cast iron. It oxidizes in the air and decom-

*To make chlorine water add dilute hydrochloric acid to a
fragment of potassium chlorate in a test tube and boil.

poses water.   Manganese forms several oxides, most important of which is manganese dioxide, MnO_2, used in making oxygen and in the manufacture of steel. Another important compound is potassium permanganate, $KMnO_4$.   Common salts of manganese are the chloride, $MnCl_2$, the nitrate, $Mn(NO_3)_2$, and the sulphate, $MnSO_4$.

Alloys. Manganese forms several alloys of importance, among which is cupro-manganese, used in making manganese bronze, manganese brass, etc.

Chief Ore. Pyrolusite, MnO_2.   Reduced by heating at a high degree with carbon.

Blowpipe Test. Fuse a small quantity of manganese dioxide on a platinum foil with plenty of sodium carbonate and potassium nitrate.   A green mass results, composed chiefly of sodium and potassium manganate, Na_2MnO_4 and K_2MnO_4.

Confirmatory Reaction. Dissolve some manganese sulphate in hot water.   Add ammonium hydroxide to alkaline reaction and then ammonium sulphide. A flesh colored precipitate of manganous sulphide, MnS, is formed.   Filter and fuse this precipitate on platinum as indicated above and obtain a green mass.

CHROMIUM. Symbol, Cr.   Combining weight, 52.1.   Specific gravity, 5.9-7.3.   Melting point, above platinum.   Chromium is a nearly infusible crystalline powder, light green in color.   It is quite uncommon in the free state, but forms many important compounds, such as potassium dichromate, $K_2Cr_2O_7$, and chrome alum, $K_2SO_4.Cr_2(SO_4)_3+24H_2O$, analogous in composition to common alum. Other important compounds, widely used as pigments, are lead chromate, $PbCrO_4$, known as *chrome yellow*, and chromic oxide, Cr_2O_3, known as *chrome green*.

Alloys. Chromium forms no alloys of importance. When added in the proportion of 0.5 to 0.75 per cent to steel it makes a hard product known as chrome steel.

Chief Ore. Chrome ironstone, $FeO.Cr_2O_3$. Metallic chromium is commonly reduced from chromic chloride by electrolysis.

Blowpipe Tests. Compounds of chromium, as chrome alum, give, when fused on platinum with sodium carbonate and potassium nitrate, a yellow mass. Boil this mass with water in an evaporating dish. Filter. Acidify the filtrate with acetic acid and add lead acetate. A yellow precipitate of lead chromate, $PbCrO_4$, is formed.

Confirmatory Reaction. Dissolve some chrome alum in water, and to a portion add ammonium hydroxide. A grayish precipitate of chromic hydroxide, $Cr_2(OH)_6$, appears. To another portion add potassium hydroxide; chromic hydroxide appears, but redissolves in excess, a compound analogous to a zincate being formed. Boil the solution and the precipitate reappears, unlike the corresponding compounds of zinc and aluminum. Filter and fuse on platinum as indicated above.

NICKEL. Symbol, Ni. Combining weight, 58.7. Specific gravity, 8.9. Melting point, 1450° C. Nickel is a highly lustrous, yellowish white metal. It is somewhat harder than iron, very brittle, but rendered so malleable by the presence of a small quantity of magnesium that it can be rolled, drawn into wire, welded, etc. It is more tenacious than iron and, like iron, it is attracted by the magnet. Nickel does not oxidize in the air at ordinary temperature, but is slowly tarnished in dry hydrogen sulphide gas, nickel sulphide, NiS, being formed. Nickel dissolves slowly in hydrochloric or sulphuric acid, but readily in nitric, forming nickel nitrate, $Ni(NO_3)_2$.

Alloys. Varieties of bronze, German silver and substitutes for German silver. An important compound of nickel and steel has lately attracted much attention as an armor plate material. Nickel is widely used in plating. It does not amalgamate directly. A nickel amalgam can be made by adding sodium amalgam to the solution of a nickel salt.

Chief Ore. Nickel blende, NiS. Reduced after roasting by heating with carbon. Nickel and cobalt occur associated in nature.

Blowpipe Tests. Metallic nickel is infusible on charcoal. Nickel compounds, as nickel nitrate, give a clear bead which is violet while hot and yellowish brown when cold.

Confirmatory Reactions. Dissolve some nickel in nitric acid. A green solution is formed. Add ammonium hydroxide and ammonium sulphide. A black precipitate of nickel sulphide, NiS, is formed, nearly insoluble in hydrochloric acid. Dissolve the precipitate by boiling in nitric acid; filter. Evaporate the solution nearly to dryness, and wash down the sides of the dish with water. Add potassium cyanide until a precipitate appears and redissolves, then an excess of sodium hypobromite and boil. A black precipitate of nickelic hydroxide, $Ni_2(OH)_6$, is formed.

COBALT. Symbol, Co. Combining weight, 59. Specific gravity, 8.9. Melting point, 1500° C. Cobalt is a white metal with a cast of red. It is harder than iron and more tenacious, but resembles this metal in malleability and ductility. It tarnishes slowly in moist air, but readily in moist hydrogen sulphide, cobalt sulphide, CoS, being formed. Like iron and nickel it is attracted by the magnet. As a metal cobalt has little application in the arts. *Smalt* is an important compound of cobalt, used as a pigment. A so-called *sympathetic ink* can be made by dissolving cobalt chloride, $CoCl_2+6H_2O$, or the nitrate, $Co(NO_3)_2+6H_2O$, in water. These solutions, when used in place of ink give an invisible line. When the paper is slightly warmed the writing becomes visible, due to the dehydration of the cobalt salt. The writing will again become invisible if allowed to absorb moisture.

Alloys. Cobalt forms few alloys of importance. Like nickel and iron, it does not amalgamate directly.

Chief Ores. Smaltite, $CoAs_2$, and cobaltite,

$CoS_2.CoAs_2$. Reduced by carbon in various ways. Its reduction is a difficult matter.

Blowpipe Test. Cobalt is infusible on charcoal. Its compounds, when fused in a borax bead made in the loop of a platinum wire, impart to the bead a deep blue color. If the bead is too strongly saturated it will appear black.

Confirmatory Reactions. Dissolve some cobalt nitrate in water. Add ammonium hydroxide and ammonium sulphide. A black precipitate of cobalt sulphide, CoS, nearly insoluble in hydrochloric acid, is formed. This precipitate gives the characteristic blue borax bead.

BARIUM. Symbol, Ba. Combining weight, 137. Specific gravity, 3.75. Melting point, 475° C. Barium is a silver white metal, ductile and malleable. It oxidizes rapidly in air or in water and never has been obtained in the coherent state. Many compounds of barium are important. Barium sulphate, $BaSO_4$, commonly known as *heavy spar*, used to adulterate paints; barium chloride, $BaCl_2$, used as a reagent in the laboratory to detect sulphuric acid and sulphates; barium dioxide, BaO_2, used in the preparation of hydrogen dioxide solutions. Barium forms no alloys. It can be made to unite with mercury by galvanic action, but the product is very unstable.

Occurrence. Heavy spar, $BaSO_4$. Metallic barium is obtained by the decomposition of the chloride by an electric current.

Blowpipe Test. Barium compounds, as barium

chloride, when held in the outer part of the Bunsen flame, on a platinum wire, impart to the flame a yellowish green color.

Confirmatory Reactions. Dissolve some barium nitrate in water. Add ammonium hydroxide and ammonium carbonate. A white precipitate of barium carbonate, $BaCO_3$, is formed. Filter, dissolve in acetic acid. Divide into two portions. To one add sulphuric acid and obtain a white, insoluble precipitate of barium sulphate, $BaSO_4$. To the second portion add potassium dichromate and obtain a yellow precipitate of barium chromate, $BaCrO_4$.

STRONTIUM. Symbol, Sr. Combining weight, 87.6. Specific gravity, 2.4. Melting point, red heat. Strontium is a slightly yellow metal, quite ductile and malleable, and harder than gold. It oxidizes rapidly in air and decomposes water. Its common compounds are strontium chloride, $SrCl_2$, strontium sulphate, $SrSO_4$, and strontium nitrate, $Sr(NO_3)_2$. The last named is used to make *red fire.* Strontium compounds are of little value in the arts. The metal does not enter into the composition of alloys. Like barium, it can be made to unite with mercury.

Occurrence. Strontianite, $SrCO_3$, and celestine, $SrSO_4$. Metallic strontium is prepared in the same manner as barium.

Blowpipe Test. Strontium compounds impart to the flame a beautiful crimson color.

Confirmatory Reaction. Dissolve some strontium

nitrate in water. Add ammonium hydroxide and ammonium carbonate. A white precipitate of strontium carbonate, $SrCO_3$, is formed. Filter, and dissolve this precipitate in acetic acid. To a portion of this solution add a weak solution of potassium sulphate and obtain a white precipitate of strontium sulphate, $SrSO_4$. To the second portion add potassium dichromate. No precipitate is formed (distinction from barium).

CALCIUM. Symbol, Ca. Combining weight, 40. Specific gravity, 1.57. Melting point, red heat. Calcium is a pale yellow metal, soft as zinc, malleable and very ductile. Calcium, like barium and strontium, has no practical application. In moist air it oxidizes, and in water it produces a violent evolution of hydrogen. Compounds of importance are calcium oxide, CaO, commonly called *lime;* calcium hydroxide, $Ca(OH)_2$, commonly called *slaked lime ;* calcium carbonate, $CaCO_3$; calcium sulphate, $CaSO_4$, known as *plaster of Paris;* calcium hypochlorite, $Ca(ClO)_2$, *bleaching powder;* and calcium chloride, $CaCl_2$. Calcium can be made to unite with mercury to form an unstable amalgam.

Occurrence. Gypsum, $CaSO_4 + 2H_2O$; limestone, $CaCO_3$. Metallic calcium is prepared in the same manner as barium.

Blowpipe Test. Compounds of calcium, as calcium chloride, impart to the flame a yellowish red color.

Confirmatory Reactions. Dissolve some calcium nitrate in water, add ammonium hydroxide and carbonate. A white precipitate of calcium carbonate, $CaCO_3$, is formed. Filter, dissolve in acetic acid and to a portion add a weak solution of potassium sulphate; no precipitate appears (distinction from strontium). Next add ammonium hydroxide and ammonium oxalate. A white precipitate of calcium oxalate, CaC_2O_4, is formed. To a second portion of the calcium solution add potassium dichromate. No precipitate is formed (distinction from barium).

MAGNESIUM. Symbol, Mg. Combining weight, 24.3. Specific gravity, 1.75. Melting point, 500° C. Magnesium is a white, hard, malleable and ductile metal. It can be rolled into ribbon and drawn into wire. It oxidizes slowly in air, but when heated above its melting point burns with a dazzling white light, the oxide, MgO, being formed. Common compounds of magnesium are magnesium oxide, MgO, commonly known as *magnesia;* magnesium carbonate, $MgCO_3$; magnesium sulphate, $MgSO_4$, commonly called *Epsom salt.* Magnesium is soluble in all common acids. It forms no alloys of importance, but can be made to form a somewhat unstable amalgam. The metal is employed to furnish the flash light in photography.

Chief Ore. Magnesite, $MgCO_3$. Found as Epsom salt in many spring waters. The metal is produced by fusing the chloride with sodium:

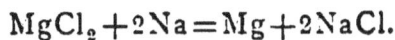

$$MgCl_2 + 2Na = Mg + 2NaCl.$$

Blowpipe Tests. Magnesium when heated on charcoal or in the Bunsen flame burns with a white light, the oxide, MgO, being formed. If this powder is dampened with cobalt nitrate and strongly ignited on charcoal it becomes pale rose in color.

Confirmatory Reactions. Dissolve a piece of magnesium in nitric acid, add ammonium hydroxide. A white precipitate appears, which redissolves in ammonium chloride. Add sodium phosphate. A white precipitate of ammonium magnesium phosphate, NH_4MgPO_4, is formed.

POTASSIUM. Symbol, K. Combining weight, 39.11. Specific gravity, 0.86. Melting point, 62.5° C. Potassium is a soft, plastic metal possessing a white metallic luster when freshly cut. It oxidizes in the air, and violently decomposes water, forming an alkaline solution, potassium hydroxide, KOH, and liberating hydrogen. For this reason it is kept under petroleum. Important compounds of potassium are the alkali just mentioned; potassium nitrate, KNO_3, commonly called *saltpeter;* potassium chlorate, $KClO_3$, both used in the manufacture of explosives; and potassium carbonate, K_2CO_3, *pearl ash.* Most potassium salts are soluble in water. Potassium unites with mercury to form an amalgam.

Occurrence. Saltpeter, KNO_3, more abundantly as potassium chloride, KCl. Reduced by fusing with carbon.

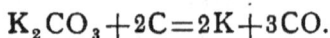

$$K_2CO_3 + 2C = 2K + 3CO.$$

Blowpipe Test. Potassium compounds, as potassium nitrate, when heated in the Bunsen flame on platinum wire and viewed through blue glass, give a violet flame. The object in using a blue glass is to shut out other flames, particularly sodium, which tend to obscure that of potassium. Potassium does not precipitate with any of the ordinary reagents.

SODIUM. Symbol, Na. Combining weight, 23. Specific gravity, .972. Melting point, 95.6° C. Sodium resembles potassium in its properties. It must be kept under oil. Important compounds of sodium are sodium hydroxide, NaOH, *caustic soda;* sodium chloride, NaCl, *common salt;* sodium carbonate, Na_2CO_3, *soda;* sodium bicarbonate, $NaHCO_3$; sodium nitrate, $NaNO_3$, *Chili saltpeter;* sodium sulphate, Na_2SO_4, *Glauber's salt;* and sodium borate, $Na_2B_4O_7 + 10H_2O$, *borax.* Anhydrous borax is known as *borax glass.* Most salts of sodium are soluble in water. Sodium forms an amalgam similar to that of potassium.

Occurrence. Common salt, NaCl, Chili saltpeter, $NaNO_3$, and in various other forms. Reduced by fusing with carbon.

$$Na_2CO_3 + 2C = 2Na + 3CO.$$

The preparation of sodium is a problem of great importance owing to its extensive use in reducing aluminum.

Blowpipe Test. Sodium compounds color the flame a bright yellow, intercepted by blue glass. Like potassium, sodium is not precipitated by any of the common reagents.

AMMONIUM. NH_4. Molecular weight, 18. Ammonium is a hypothetical substance, not known in the free state but supposed to exist in the so-called ammonium salts and there perform the part of a metal. The support for the theory of the existence of this substance lies in the following facts: When sodium amalgam is thrown into a concentrated solution of ammonium chloride a spongy, metallic mass rises to the surface of the liquid. This is commonly called ammonium amalgam; and it is generally supposed that in this amalgam the ammonia gas and hydrogen exist united as NH_4, for when an attempt is made to separate the mercury by heat the substance decomposes into ammonia gas, NH_3, hydrogen and mercury. Compounds of ammonium are ammonium hydroxide, NH_4OH, commonly called *ammonia water;* ammonium chloride, NH_4Cl, often called *sal ammoniac;* ammonium nitrate, NH_4NO_3; ammonium carbonate, $(NH_4)_2CO_3$; ammonium sulphide, $(NH_4)_2S$, and ammonium oxalate, $(NH_4)_2C_2O_4$. Nearly all ammonium salts are soluble in water, and are volatilized by heat, in some cases with decomposition. In most of their chemical properties these salts are analogous to the salts of the alkalies proper, the group (NH_4) acting as a single metal, for example, as K or Na.

Blowpipe Tests. The common salts, as ammonium chloride, when heated on charcoal or in a test tube vaporize, undecomposed, yielding dense white fumes. This is not true of ammonium nitrate, which when heated gives off water and nitrous oxide, N_2O, commonly called *laughing gas.*

Confirmatory Reactions. Ammonia gas is recognized by its odor and by its power of turning red litmus paper blue. Note, however, that ammonium chloride possesses no odor. Dissolve some in water, add a strong solution of potassium hydroxide and boil. Ammonia gas, NH_3, is liberated and can be detected by its odor or by its action on moist red litmus paper.

GOLD. Symbol, Au. Combining weight, 197. Specific gravity, 19.26-19.31. Melting point, 1075° C. Gold is a yellow, lustrous metal not affected by air, by moisture or by sulphur and its compounds—properties which have made it of great use for ornaments, from the remotest time. The color of gold is often modified, however; in the molten state it is green, when precipitated from a solution it is often brown, becoming yellow again when fused, and in the form of foil it transmits a green light. In malleability and ductility gold surpasses all other metals. It can be hammered into foil $\frac{1}{300000}$ of an inch in thickness and drawn into wire so fine that one mile of it will weigh less than a gram. Gold is nearly as soft as lead and about as tenacious as silver. Owing to its softness in the pure state, it must be alloyed to give it sufficient hardness to resist wear when used for coinage or jewelry. In alloying, copper and silver are used. Copper makes gold reddish in color, harder, more fusible and less ductile and malleable. Silver hardens gold, affects its malleability less than any other metal, but greatly modifies its color. The addition of five per cent of silver makes an appreciable difference in the color, while fifty per cent destroys it. Varieties of gold alloyed with varying proportions of silver bear the names of *yellow gold*, *green gold* and *pale gold*. Alloys other than those mentioned are of little value. The presence of small quantities of lead, bismuth or antimony in gold makes it very brittle and greatly affects its color and luster. As little as 0.05 per cent of lead or of antimony impairs its malleability and slightly greater quantities render it unworkable. From the facts just stated it is apparent that in working gold great care must be exercised to prevent it from becoming contaminated with base metals, particularly those mentioned. Thus after annealing, while the metal is still hot, it never should be placed upon a lead table top or brought in contact with any metals which are liable to alloy with it at comparatively low temperatures. After gold plates are swaged they should be carefully cleaned to remove adhering particles of die metal which would diffuse into them upon annealing. This can be done by "pickling" in dilute nitric acid,* or by polishing with pumice stone on a brush-wheel.

In the preparation of gold foil perfectly pure gold is usually used. It is first rolled into ribbon about $\frac{1}{300}$ of an inch in thickness, then cut into pieces an inch square and hammered between sheets of tough paper or vellum, and finally animal membrane,

---

*An error often made by the student consists in "pickling" the plate in sulphuric acid to remove bits of die metal. It should be remembered that lead, antimony and certain other metals commonly employed in alloys for dies are insoluble in this acid.

termed gold-beater's skin, until it is about $\frac{1}{200000}$ of an inch in thickness. These sheets of gold are next cut and made into books. Gold foil is used in dentistry for filling teeth. Two varieties, the *cohesive* and the *noncohesive*, are commonly employed. The cohesive is simply pure gold which, when firmly packed in the cavity of the tooth by means of pluggers, becomes welded owing to its characteristic cohesive property. The cohesiveness of this variety of gold is greatly diminished by the accumulation of moisture and gases upon its surface. By properly annealing, however, in the flame of a spirit lamp, in a muffle, or preferably upon a sheet of mica or metal, this property can be almost entirely restored. In the case of noncohesive gold union does not take place between the different pieces. This result is probably attained by slightly alloying or by subjecting the pure foil to the action of some gaseous substance which becomes deposited on its surface. The process of manufacturing noncohesive gold is a trade secret.

Gold is insoluble in hydrochloric, sulphuric or nitric acid, but soluble in potassium cyanide, in free chlorine and in aqua regia, gold chloride, $AuCl_3$, being formed in the last two cases.

Alloys. Gold coin, gold plate, gold solders, etc. Gold unites very readily with mercury. The fineness of gold is usually expressed in carats. Pure gold is twenty-four carats, or 1000 fine; twenty-two carat gold contains twenty-two parts of pure gold and two of other metal, and may be expressed as 916.6 fine. When a gold-silver or gold-copper alloy contains less than twenty-five per cent of gold, nitric acid will disintegrate it; but when the gold is over this amount, sufficient silver or copper to bring the proportion of gold down to that stated must be fused with it before the acid will completely remove the alloying metal and leave the gold in a pure state. This is known as *quartation*.

Chief Ores. Gold is usually found in the metallic state, often as nodules or nuggets. Its chief sources are quartz veins and alluvial deposits. In the first case the ore is crushed into a fine powder and the gold separated by amalgamation. Subsequently the amalgam is placed in iron retorts and the mercury is removed by distillation. Mining of this kind is commonly called *vein mining*. In the second case the gold is washed with water in a "cradle," or pan. The gold being heavy remains mixed with sand and gravel, while the lighter substances are washed away. It then may be further separated from impurities by amalgamation, as indicated above. This is commonly known as *placer mining*. A modification of placer mining is *hydraulic mining*. It consists in directing a stream of water under high pressure against the mountain side, thus washing down and carrying away the loose gold-bearing earth and depositing the gold in sluices.

After obtaining the gold by any of the foregoing processes it must be refined.* This may be done in several ways, for example by quartation or by dis-

---

*See Chapter XII.

solving in aqua regia and precipitating the gold by some reducing agent, as ferrous sulphate.

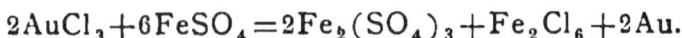

$$2AuCl_3 + 6FeSO_4 = 2Fe_2(SO_4)_3 + Fe_2Cl_6 + 2Au.$$

Other processes of extracting gold from its ores are *chlorination* and *cyanide*. In both processes the ore is crushed, oxidized in a reverberatory furnace and placed in large wooden vats. It is then treated with chlorine gas or leached with a solution of potassium cyanide. In the former case the gold is converted into the chloride, and after being dissolved in water is treated with ferrous sulphate to precipitate the gold as shown above. In the latter case the gold is dissolved by the cyanide solution and then precipitated by means of an electric current.

Blowpipe Tests. Gold remains a bright yellow after fusing on charcoal. In the melted state the bead appears green, and if allowed to cool undisturbed, first turns dark on the surface, then suddenly "flashes," becomes yellow and solidifies. This is more noticeable with large masses.

Confirmatory Reactions. When a solution of stannous and stannic chlorides is added to gold chloride, $AuCl_3$, a purple precipitate of unknown composition, commonly called *purple of Cassius*, is formed. Upon adding a solution of ferrous sulphate to gold chloride a dark brown precipitate of metallic gold results.

PLATINUM. Symbol, Pt. Combining weight, 195. Specific gravity, 21.5. Melting point, 2000° C. Platinum is a tin white metal, not affected by air, moisture or hydrogen sulphide. It is infusible except in the oxyhydrogen flame, and not acted upon but by few substances; hence its wide use in the arts and sciences, particularly in chemistry, for crucibles, wire, etc. When made into crucibles, etc., it is usually alloyed with iridium, thus increasing its hardness, melting point and resistance to reagents. Like iron, it can be welded at white heat, and by means of the oxyhydrogen blowpipe it can be soldered autogenously, i. e., with itself. Platinum is more tenacious than silver, about as malleable as tin, and more ductile than other metals except gold and silver. Platinum wire can be drawn to $\frac{1}{1200}$ of an inch in diameter. In the melted state platinum, like silver, absorbs oxygen, which in being expelled causes " spitting." Finely divided platinum, particularly *spongy platinum*, possesses the peculiar property of condensing gases upon its surface. It can absorb many times its own volume of oxygen, hydrogen and other gases. Spongy platinum is made by igniting ammonium platinic chloride. *Platinum black* is a variety of finely divided platinum, often used in dentistry to color the enamel of artificial teeth. It may be prepared by dissolving platinum in aqua regia, evaporating the excess of acid, diluting with water and adding this solution to a boiling solution of glycerine and potassium hydroxide, or by boiling platinic chloride with a strong solution of potassium hydroxide and adding grape sugar. Platinum is insoluble in the common acids and dissolves more slowly than gold in aqua regia, forming platinic chloride, $PtCl_4$.

Alloys. Platinum is used, alloyed with rare metals, as indicated above. It enters somewhat into the composition of dental alloys. It unites with most metals, but forms no combination, when compact, with mercury. Platinum sponge, however, forms an amalgam when triturated in a warm mortar with mercury and acetic acid. A small proportion of platinum in gold renders this metal very elastic.

Chief Ore. Platinum is found in the metallic form alloyed with certain rare metals, as palladium, rhodium, iridium, osmium, etc. The metal is not usually separated completely from these metals, but made into a pure alloy with them by fusing in a furnace heated by an oxyhydrogen blowpipe.

When pure platinum is desired, the ore is first treated with nitric acid to dissolve any copper or iron present and then with dilute aqua regia. The platinum dissolves and is precipitated from this solution as ammonium platinic chloride by ammonium chloride. Upon being heated, this compound is decomposed and the finely divided platinum resulting is converted into the compact variety by mixing with water and molding under high pressure and finally welding at a white heat.

Blowpipe Test. Platinum is infusible on charcoal with the ordinary blowpipe flame.

Confirmatory Reactions. Platinum dissolves slowly in aqua regia and is precipitated from this solution by alcohol and ammonium chloride as ammonium platinic chloride, $(NH_4)_2PtCl_6$, a yellow, crystalline precipitate.

CHAPTER III.

QUALITATIVE CHEMICAL ANALYSIS.

Having become acquainted with some of the physical and chemical tests which serve to distinguish one metal from another, the student next proceeds to apply these tests so that the metals, when existing in the form of complex mixtures, may be separated and identified. As already seen, it is not a difficult matter to distinguish between the metals when existing alone either in the metallic state or in solution; thus metallic silver or a silver solution is easily distinguished from the corresponding forms of lead. If, however, these metals exist in an alloy or in a complex solution the tests outlined in Chapter II. must be applied in a certain definite order. Unfortunately, each metal does not have a test solution or *reagent* of its own that can be added to a complex substance and identify that metal under all conditions. On the contrary, the addition of a reagent may be responded to by several metals, and it follows that the test for a metal is free from uncertainty only when it is definitely known that other metals responding to the same test are absent. It is evident, then, that *qualitative analysis*, far from being an application of tests without regard to order, is a systematic scheme whereby a metal is separated in such a manner as to leave no doubt concerning its identity. Further, to illustrate, in the *Confirmatory Reactions* of the metals in Chapter II., it was observed that a silver solution gave, with hydrochloric acid, a white precipitate of silver chloride and this was called a test for silver. It was further observed, however, that lead and certain mercury solutions gave a similar white precipitate with the same reagent; and for this reason the test for silver is not final unless it is definitely known that these metals are absent. Beyond the fact, however, that these precipitates are alike in color, and are produced by the same reagent, they do not resemble one another. Lead chloride dissolves in hot water while the chlorides of silver and mercury do not. Silver chloride dissolves readily in ammonium hydroxide while mercurous chloride blackens. Thus is offered a method of positively identifying these metals, whether alone or in complex mixture.

In the case just cited hydrochloric acid did not precipitate the metals in the metallic form but as chlorides, insoluble compounds of these metals. This is uniformly the case in qualitative analysis. There are few cases in which metals are precipitated as such. The reagent generally forms with the metal in solution a compound insoluble in the menstruum, hence it precipitates. As a rule in qualitative analysis, the substance, if not already in solution, is dissolved by some solvent, as water or acid, before the tests are applied.

GROUPING OF THE METALS.

As seen in the foregoing illustration, silver, lead and mercury are precipitated from their solutions as chlorides by means of hydrochloric acid. On this fact is based a method for separating these metals from all others, as they are the only ones forming insoluble precipitates with this acid. If, then, hydrochloric acid is added to a mixture containing a great number of metals and a white precipitate appears, it is evident that some of these metals are present. On adding sufficient acid they will be completely precipitated and can be removed from the solution by filtration. If to the filtrate from which these metals have been removed hydrogen sulphide is added, another precipitate may appear, which may contain the following metals, viz., arsenic, antimony, tin, bismuth, copper, cadmium and some lead and mercury. This, again, can be separated from the solution by filtration. Hence, by the successive application of reagents capable of removing from the original solution certain metals, all are separated into *groups*. In this manner the analysis of a complex solution is somewhat simplified, as it is known that each group contains only a limited number of metals. It next remains to submit each group to a special series of tests in order to determine the individual metals present. Reagents like hydrochloric acid and hydrogen sulphide are commonly called *group reagents*. In the following table are given the five groups of metals and their corresponding group reagents. It should always be borne in mind that in actual analysis each group is removed in its order; thus hydrochloric acid is first added to the solution, the resulting precipitate removed by filtration, then to the filtrate hydrogen sulphide is added, etc. If a group reagent fails to produce a precipitate, no metals of that group are present, and the solution is taken to the next group.

GROUP TABLE.

Group Reagents.

From the original solution HCl precipitates
- Lead as $PbCl_2$ (white)
- Silver as $AgCl$ (white)
- Mercury (ous) Hg_2Cl_2 (white)

Group I.

From Group I. filtrate H_2S precipitates
- Arsenic as As_2S_3 (yellow)
- Antimony as Sb_2S_3 (orange red)
- Tin as SnS or SnS_2 (brown or yellow)
- Bismuth as Bi_2S_3 (black)   [low]
- Copper as CuS (black)
- Cadmium as CdS (yellow)
- Mercury (ic) as HgS (black)
- Lead as PbS (black)

Group II.

From Group II. filtrate NH_4OH and $(NH_4)_2S$ precipitate
- Iron as FeS (black)
- Aluminum as $Al_2(OH)_6$ (white)
- Chromium as $Cr_2(OH)_6$ (green)
- Manganese as MnS (flesh color)
- Zinc as ZnS (white)
- Nickel as NiS (black)
- Cobalt as CoS (black)

Group III.

From Group III. filtrate $(NH_4)_2CO_3$ precipitates
- Barium as $BaCO_3$ (white)
- Strontium as $SrSO_3$ (white)
- Calcium as $CaCO_3$ (white)

Group IV.

Group IV. filtrate will contain
- Magnesium
- Sodium
- Potassium
- Ammonium

Group V.

NOTE.—Separate portions of the Group IV. filtrate are used in testing for Mg, Na and K. NH_4 is looked for in the original solution.

CHAPTER IV.*

ANALYSIS OF GROUP I.

Separation of Lead, Silver and Mercury.

To the unknown solution add hydrochloric acid, stirring vigorously, in sufficient quantity to completely precipitate Group I. This is best determined by adding the reagent as long as a precipitate seems to be formed, and then filtering. If a drop more of acid produces further precipitation, add more and filter again through another paper. *Keep the filtrate for Group II.* Wash the precipitate with cold, distilled water. *Always reject washings unless directed to the contrary.* Heat some water in a test tube and pour over the precipitate on the filter. Collect the water in a second test tube as it passes through. Reheat, pour over again and collect as before.

Lead chloride is soluble in hot water and if present in the precipitate will be removed and can be tested for in the water solution by adding sulphuric acid. Lead salts give with sulphuric acid a fine white precipitate of lead sulphate, $PbSO_4$. To the remainder of the precipitate on the filter, which

---

*In order fully to understand the reactions employed in qualitative analysis consult the *Confirmatory Reactions* given under each metal in Chapter II.

may contain the chlorides of silver and mercury(ous), add ammonium hydroxide. Silver chloride is soluble in the latter and can be tested for in the ammoniacal solution by acidifying with nitric acid. The presence of silver is indicated by the appearance of a white, curdy precipitate of silver chloride, AgCl. Finally, if mercurous chloride is present in the precipitate it will remain on the filter paper as a black residue, mercurous ammonium chloride, NH_2Hg_2Cl, after the above treatment with ammonium hydroxide. This black residue is sufficient evidence of the presence of mercury.

After following these directions until familiar with the details, it will be found more convenient to use Table I. on page 80, which is an abbreviated outline of the process just given.*

---

*The form of report to be submitted by the student after completing the analysis of an unknown substance is shown in the Appendix, Section II.

CHAPTER V.

ANALYSIS OF GROUP II.

Separation of Arsenic, Antimony, Tin, Bismuth, Copper, Cadmium, Mercury and Lead.

To the *filtrate from Group I.* add hydrogen sulphide gas as long as a precipitate seems to form. Filter and pass more gas into the filtrate. Under certain conditions some of the metals of this group are slow in precipitating. If a precipitate again appears, filter and *keep the filtrate for Group III.* Wash the precipitate once with water and proceed to determine the metals. The color of the precipitate often gives some idea as to its composition. The sulphides of lead,* mercury,† bismuth and copper are black, those of cadmium and arsenic, light yellow, that of antimony, brick red, and that of tin, dark brown. Remove precipitate and paper from the funnel, place in a porcelain evaporating dish and cover with yellow ammonium sulphide. Heat gently, i. e., "digest" for some time and filter. The sulphides

---

*If lead was found in Group I. it will usually appear here owing to the slight solubility of lead chloride in cold water and in hydrochloric acid.

†The mercury referred to here is a mercuric compound. Mercuric salts, unlike mercurous, are not precipitated by hydrochloric acid in Group I.

of arsenic, antimony and tin are soluble in ammonium sulphide, and will now be found in this filtrate, while the balance of the sulphides remain as a residue on the filter paper. Proceed as follows with the ammonium sulphide solution : To it add hydrochloric acid. If any of the three soluble sulphides are present they appear as a more or less highly colored precipitate. Filter and reject the filtrate. Redissolve this precipitate by digesting in a porcelain dish with chlorine water.

As more or less sulphur is always present here, it will remain as a white or colored insoluble residue. Filter out any sulphur, boil to expel chlorine, and test for arsenic, antimony and tin as follows : In a hydrogen generator (see Fig. 25) place a few pieces of zinc and cover with hydrochloric acid. It is obvious that zinc free from arsenic must be used. When the hydrogen has run for a short time, pour the solution to be tested for arsenic, antimony and tin into the generator through the funnel tube. Allow the gas from the generator to bubble into some silver nitrate solution in a test tube for five or ten minutes. Arsenic and antimony are converted by the nascent hydrogen into gaseous compounds, which pass into the silver nitrate, while the tin remains in the generator. A black precipitate in the silver nitrate indicates arsenic or antimony. Confirm by filtering and treating as follows : (Save the filtrate, as it may contain arsenic.) Wash the precipitate very thoroughly with water, digest in warm potassium hydroxide, filter, acidify with hydrochloric acid and add hydrogen sulphide. An orange red precipitate, antimony sulphide, Sb_2S_3, indicates antimony. To the above filtrate, which may contain arsenic, add hydrochloric acid to remove the silver. Filter, reject the precipitate, which is silver chloride, and test the filtrate for arsenic by hydrogen sulphide. A yellow precipitate, arsenious sulphide, As_2S_3, indicates arsenic. Next turn to the generator and test its contents for tin. If the zinc is not entirely dissolved, add a little concentrated hydrochloric acid, and, when dissolved, filter a portion of the liquid in the generator and test it for tin by adding mercuric chloride. A white precipitate, mercurous chloride, Hg_2Cl_2, indicates tin. Return now to the residue which remained after treating the Group II. precipitate with yellow ammonium sulphide. This residue may contain the sulphides of mercury, lead, bismuth, copper and cadmium. Place it, together with the filter paper, in a porcelain dish and digest with nitric acid. Everything goes into solution except mercuric sulphide, which remains as a black residue. Filter, and treat the filtrate as indicated below. Test the black residue for mercury by dissolving in aqua regia, boiling and adding stannous chloride. A white precipitate, mercurous chloride, Hg_2Cl_2, is final evidence of mercury. To the filtrate, which may contain the nitrates of lead, bismuth, copper and cadmium, add a few drops of sulphuric acid and evaporate nearly to dryness. Dilute with water in a beaker. A white precipitate, lead sulphate, $PbSO_4$, indicates lead. Filter, and to the filtrate add ammonium hydroxide in excess of alkalinity. If

copper is present the solution will assume a deep
blue color, sufficient evidence of copper, and in this
solution will be distinguished a white, gelatinous precipitate, bismuth hydroxide, $Bi(OH)_3$, if bismuth is
present. Further confirm by filtering and testing as
follows : (Keep the filtrate to test for cadmium.) To
the precipitate on the filter paper, supposed to be
bismuth hydroxide, add *stannite** and if bismuth, it
will immediately turn black. To the blue filtrate
which remains to be tested for cadmium add potassium
cyanide until the color disappears and then pass in
hydrogen sulphide gas. If cadmium is present a
yellow precipitate of cadmium sulphide, CdS, appears.
It should be noted, however, that if copper is not
present, indicated by a colorless solution, it is unnecessary to add potassium cyanide, but the filtrate
from the precipitate of bismuth hydroxide can be
tested at once for cadmium by passing into it hydrogen sulphide gas. The purpose of adding potassium
cyanide to the blue solution is to convert the copper into a form which does not precipitate with
hydrogen sulphide, else the black copper sulphide
formed would obscure the yellow of the cadmium
sulphide. (Hereafter substitute Table II.)

See note, page 30.

CHAPTER VI.

ANALYSIS OF GROUP III.*

Separation of Zinc, Aluminum, Iron, Manganese, Chromium, Nickel and Cobalt.

To the *filtrate from Group II.* add a few drops of ammonium chloride, then ammonium hydroxide to alkaline reaction. The purpose of adding ammonium chloride is to keep magnesium, a fifth group metal, from precipitating here as an hydroxide. After the addition of ammonium hydroxide, a gelatinous precipitate of the hydroxides of iron, aluminum and chromium may appear. Disregard this and add enough ammonium sulphide to make the precipitation complete. Filter, and add to the filtrate a few more drops of ammonium sulphide to insure complete precipitation. *Keep the filtrate for Group IV.* Treat the precipitate as follows: Wash once

---

*In case phosphates are present in the solution, a modification of this method must be employed, owing to the fact that barium, strontium, calcium and magnesium precipitate as phosphates upon the addition of ammonium hydroxide. Phosphoric acid is detected by boiling a small portion of Group II filtrate first, to remove hydrogen sulphide, then with a drop of nitric acid and an excess of ammonium molybdate. A yellow precipitate is ammonium phospho-molybdate. Should phosphoric acid be found by the test just suggested, treat the filtrate from Group II. according to Table VI. instead of Table III.

with water and then pour cold hydrochloric acid over
it, returning the same portion of acid to the filter two
or three times. A black residue, insoluble in hydrochloric acid, indicates nickel or cobalt. Test this
residue as indicated later. Evaporate the hydrochloric acid solution, which may contain zinc, aluminum, chromium, iron and manganese, nearly to dryness in a porcelain dish. Wash down the sides of
the dish with water and add an excess of potassium
hydroxide. Boil for a minute and filter. Iron, manganese, chromium and some nickel and cobalt are precipitated as hydroxides, while zinc and aluminum are
precipitated but redissolved by the alkali and will be
found in the filtrate. Test the filtrate for zinc and
aluminum. Divide into two portions. To one add
ammonium sulphide. A white precipitate, zinc sulphide, ZnS, indicates zinc. To the second portion
add twice its volume of ammonium chloride, boil and
let stand a few minutes. A white, flaky precipitate,
aluminum hydroxide, $Al_2(OH)_6$, is evidence of aluminum. The precipitate obtained above with potassium
hydroxide, test for iron, manganese and chromium
as follows: Remove a small portion from the filter
paper with a glass rod. Place it in a test tube and dissolve in chlorine water. Boil to expel free chlorine
and test for iron by adding potassium sulphocyanate,
KCNS. A red coloration, $Fe_2(CNS)_6$, is evidence of
iron. Remove the remainder of the precipitate and
the paper from the funnel, roll it up into a small
mass, place it on a platinum foil and fuse with twice
its bulk of potassium nitrate and sodium carbonate.

Make this fusion perfect, using the blowpipe if necessary. After cooling, the mass will be dark green in
color if manganese is present, due to the formation
of a manganate, K_2MnO_4. Dissolve the mass in
boiling water. Filter, acidify with acetic acid, boil
and add lead acetate. A yellow precipitate of lead
chromate, $PbCrO_4$, will appear if chromium is present.
A white precipitate will sometimes appear here due
to the insufficient addition of acetic acid. This is no
evidence of chromium. Return now to the black residue, which was insoluble in hydrochloric acid. Test
it for cobalt first by means of the borax bead. A
blue bead proves cobalt. To test for nickel dissolve
the residue in dilute nitric acid by digesting in a
porcelain dish. Filter and evaporate the filtrate
nearly to dryness. Take up with a little water, add
potassium cyanide until a precipitate forms and redissolves, then an excess of sodium hypobromite. If
nickel is present it shows itself as a dense black precipitate of nickelic hydroxide, $Ni_2(OH)_6$. (Hereafter
substitute Table III.)

CHAPTER VII.

ANALYSIS OF GROUP IV.

Separation of Barium, Strontium and Calcium.

To the *filtrate from Group III.* add ammonium carbonate. Warm and filter. Add a little more of the reagent to insure complete precipitation. If no further precipitate is formed *keep the filtrate for Group V.* Wash the precipitate of carbonates with water and dissolve in acetic acid, using very little and pouring it on several times. Everything dissolves. Add to this acetic acid solution an excess of potassium dichromate. A yellow precipitate of barium chromate, $BaCrO_4$, appears if barium is present in the solution. Warm and filter. The filtrate can now be tested for strontium and calcium. Potassium dichromate being a colored reagent, the filtrate is always yellow if it is in excess, and the latter condition is essential to insure complete precipitation. It is well to first separate any calcium or strontium from the dichromate solution. This is done by adding ammonium hydroxide until alkaline, and then reprecipitating strontium and calcium with ammonium carbonate. Filter, wash thoroughly to remove the colored liquid from the precipitate and reject both washings and filtrate. Redissolve the precipitate with acetic acid as before, add a weak solution of potassium sulphate and allow the liquid to stand for ten minutes; a white precipitate is strontium sulphate, $SrSO_4$. Filter. To the filtrate add ammonium hydroxide and ammonium oxalate. If calcium is present a fine, white precipitate of cal- cium oxalate, CaC_2O_4, will form. (Hereafter sub- stitute Table IV.)

CHAPTER VIII.

ANALYSIS OF GROUP V.

Testing for Magnesium, Sodium, Potassium and Ammonium.

With the exception of ammonium, the metals of this group should be looked for in the filtrate from Group IV. To test for magnesium take a small portion of Group IV. filtrate, add ammonium chloride and hydroxide, and then sodium phosphate. A white precipitate, ammonium magnesium phosphate, NH_4MgPO_4, appears after a time if magnesium is present. Into another portion of the filtrate from Group IV. dip a clean platinum wire and hold in the flame. If sodium is present the flame above the wire will be colored a bright yellow. Again dip the wire into the solution, hold it in the flame and observe the flame through a piece of blue glass. A violet flame indicates potassium. The student should become familiar with the potassium flame by testing some solution known to contain potassium, as potassium sulphate. The object of using the blue glass, as already explained, is to shut out any sodium color, which is usually so bright as to obscure small quantities of potassium.

It is evident that ammonium cannot be looked for here, as its compounds have been employed as reagent several times during the course of the analysis. Therefore, take a portion of the *original solution* in a test tube, add potassium hydroxide to strong alkaline reaction; a precipitate may occur, but disregard it, and boil the contents of the test tube vigorously. Ammonium compounds are decomposed by the alkali and ammonia gas is liberated. This may be detected by its odor or by its action on damp red litmus paper. (Hereafter substitute Table V.)

CHAPTER IX.

TREATMENT OF METALS AND ALLOYS.

In previous chapters the analysis of complex mixtures of metals in solution was considered. It often occurs in practical analysis, however, that instead of a mixture of metallic salts in solution the substance to be analyzed is a metal or a mixture of metals, such as an alloy.

The analysis of such a substance usually consists in converting it into solution by some acid and then applying the ordinary qualitative scheme beginning at Group I. It is obvious that in this work such metals as barium, strontium, calcium and the alkali metals need not be looked for, as they do not enter into the composition of alloys.

In every case a careful examination of the physical properties, such as color, crystalline form, fusibility, etc., should precede an attempt to convert the substance into solution. Often the blowpipe reactions serve to identify the substance, at least in part. Thus, if a small piece easily fuses into a globule on charcoal, any of the fusible metals referred to in Chapter II. may be present. On the other hand, if the substance is infusible, iron, cobalt, nickel or platinum may be present. Further, the incrustation obtained may throw some light upon the presence of certain oxidizable metals. The next step is to convert the substance into solution. It was observed in Chapter II. that all the metals are not soluble in any single acid but that nitric acid is the most common solvent; hence nitric acid is generally used in this connection, as it dissolves all the metals except tin, antimony, gold and platinum.

METHOD.

Take a small quantity of the metal, preferably in the form of filings or clippings, and treat with strong nitric acid in an evaporating dish. Apply heat gently, adding small portions of acid occasionally to overcome the loss by evaporation until the substance is wholly dissolved or disintegrated.

I. If the substance has completely dissolved, evaporate the solution almost to dryness to remove the excess of acid. Add considerable water and proceed as in the analysis of solutions, starting with Group I. Should the solution become turbid upon diluting with water bismuth is likely present. Add enough nitric acid to remove this turbidity.

II. If the substance has not completely dissolved in nitric acid, pour off the clear liquid, add more acid, boil and again pour off. Treat the clear solution as indicated in I. Wash the residue from the evaporating dish into a filter and then wash thoroughly with water. This insoluble residue may contain gold, platinum, tin or antimony, the last two in the form of metastannic and antimonic acids. Boil with concentrated hydrochloric acid in an evaporating dish for ten minutes, then add water and boil. Filter, and treat the filtrate for tin and antimony as indicated in Chapter V. If a dark residue remains after boiling with hydrochloric acid, gold or platinum may be looked for.

Residue. Digest in aqua regia until dissolved. Gold and platinic chlorides are formed. Dilute with a little water and filter if not perfectly clear. Divide into two portions.

1. Test one for gold. Dilute considerably and add a clear solution of ferrous sulphate. A brown precipitate of metallic gold will be formed if gold is present.

2. Test the second portion for platinum by evaporating nearly to dryness and adding ammonium chloride and alcohol. After standing some time a yellow precipitate of ammonium platinic chloride will separate if platinum is present.

TABLE I.

ANALYSIS OF GROUP I.

Separation of Lead, Silver and Mercury.

To the *solution* add *hydrochloric acid.* Stir and filter. Treat the precipitate as indicated below. *Take the filtrate to Table II.*

Wash the precipitate with cold water. Treat with *hot water.*

Hot Water Solution may contain **lead.** Test by H_2SO_4. A precipitate is $PbSO_4$ (**white**).	Remaining Precipitate on filter may contain **silver** and **mercury.** Treat with **ammonium hydroxide.**	
	Ammonium Hydroxide Solution may contain **silver.** Test by acidifying with HNO_3. A curdy precipitate is AgCl (**white**).	**Remaining Precipitate** on filter, if black, contains **mercury,** due to the formation of mercurous ammonium chloride, NH_2Hg_2Cl (**black**).

TABLE II.

ANALYSIS OF GROUP II.

Separation of Arsenic, Antimony, Tin, Bismuth, Copper, Cadmium, Lead and Mercury.

To the *filtrate from Group I.* add *hydrogen sulphide gas.* Warm and filter. Treat the *precipitate* as indicated below. *Take the filtrate to Table III.*

Wash the precipitate and digest it in an evaporating dish with *yellow ammonium sulphide.* Filter. Treat the *residue* for *bismuth, copper, cadmium, lead* and *mercury* as indicated in "B." To the *ammonium sulphide solution* add *hydrochloric acid*; filter and wash. (Reject the filtrate.) Examine the *precipitate* according to "A." for *arsenic, antimony* and *tin.*

"A."

Digest the *precipitate* in an evaporating dish with *chlorine water.* Filter out any sulphur, boil to remove chlorine, pour into a *hydrogen generator.* Generate hydrogen with *zinc* and *hydrochloric acid* and allow the gas to pass into *silver nitrate* solution for five or ten minutes. Look for *tin* in the *generator.* Filter and examine the *precipitate for antimony* and the *filtrate for arsenic.*

Precipitate may contain **antimony.** Treat on filter with warm **potassium hydroxide,** filter, acidify with **hydrochloric acid** and add **H$_2$S.** A precipitate is Sb$_2$S$_3$ (**orange red**).	**Filtrate** (silver nitrate solution) may contain **arsenic.** Remove the silver by adding **hydrochloric acid** and filtering. To the filtrate add **H$_2$S.** A precipitate is As$_2$S$_3$ (**yellow**).	**Hydrogen Generator** may contain **tin.** Add more hydrochloric acid, if necessary, and when all the zinc is dissolved, filter a portion of the liquid in the generator, and add **HgCl$_2$.** A precipitate, Hg$_2$Cl$_2$, (**white**) is evidence of tin.

"B."

Digest the *residue*, not soluble in yellow ammonium sulphide, in an evaporating dish with *nitric acid.* Filter.

Black Residue insoluble in nitric acid indicates **mercury.** To confirm: Dissolve in **aqua regia,** boil and add $SnCl_2$. A precipitate is Hg_2Cl_2 (**white**).	**Filtrate** may contain **lead, bismuth, copper** and **cadmium.** Add **sulphuric acid** and evaporate nearly to dryness. Dilute with water. If a white precipitate appears, filter.

Precipitate indicates **lead.** $PbSO_4$ (**white**).	**Filtrate** may contain **bismuth, copper** and **cadmium.** Add **ammonium hydroxide** to alkaline reaction. Filter.	
	Precipitate indicates **bismuth,** $Bi(OH)_3$ (**white**). Confirm by pouring hot **stannite*** over it on the filter. Precipitate turns black.	**Filtrate:** If **blue** contains **copper** and may contain **cadmium.** To the blue solution add **potassium cyanide** until colorless, then H_2S. A precipitate is CdS (**yellow**).

*See note, page 30.

TABLE III.*

ANALYSIS OF GROUP III.

Separation of Zinc, Aluminum, Iron, Manganese, Chromium, Nickel and Cobalt.

To the *filtrate from Group II.* add a little *ammonium chloride*, then *ammonium hydroxide*, till alkaline, and *ammonium sulphide*. Filter, wash and treat the *precipitate* as indicated below. *Take the filtrate to Table IV.*

Pour *hydrochloric acid* over the *precipitate* on the filter. Examine any insoluble *black residue* for nickel and cobalt. (See pages 44-46.) Evaporate the hydrochloric acid solution nearly to dryness in an evaporating dish, take up with water, add excess of *potassium hydroxide* and boil. Filter.

Filtrate may contain **zinc** and **aluminum.** Divide into two portions.	**Precipitate** may contain **iron, manganese, chromium** and some **nickel** and **cobalt.** Test a portion of the **precipitate** for iron and another for manganese and chromium.

Zinc.	Aluminum.	Iron.	Manganese and Chromium.
To one portion add $(NH_4)_2S$. A precipitate is ZnS (white).	To the second portion add NH_4Cl and boil. A precipitate is $Al_2(OH)_6$ (white).	Dissolve a portion of the **precipitate** in a test tube in **chlorine water.** Boil to expel free chlorine and add **KCNS. A red solution** is $Fe_2(CNS)_6$.	Fuse a large portion of the **precipitate** on a platinum foil with **sodium carbonate** and **potassium nitrate. A green mass** is K_2MnO_4. Dissolve the green mass in boiling water, filter, acidify the filtrate with acetic acid and add $Pb(C_2H_3O_2)_2$. A precipitate is $PbCrO_4$ (yellow).

*This table can be used only when phosphates are absent.

TABLE IV.

ANALYSIS OF GROUP IV.

Separation of Barium, Strontium and Calcium.

To the *filtrate from Group III.* add *ammonium carbonate.* Warm and filter. Treat the well washed precipitate as indicated below. *Take the filtrate to Table V.*

Dissolve the *precipitate* on the filter with *acetic acid.* To the solution add *potassium dichromate* in excess. Warm and filter.

Precipitate indicates **barium**, BaCrO₄ (yellow).	**Filtrate (yellow)** may contain **strontium** and **calcium.** Add **ammonium hydroxide** till alkaline, then **ammonium carbonate.** Filter and reject the filtrate. Wash the **precipitate** and paper until yellow color is removed. Dissolve the **precipitate** in the filter with **acetic acid.** To this acetate solution add dilute **potassium sulphate**; let stand and filter.

Precipitate indicates **strontium**, SrSO₄ (white). Confirm by flame test. (See Blowpipe Test, page 47.)	**Filtrate** may contain **calcium.** Add **ammonium hydroxide** till alkaline, then (NH₄)₂C₂O₄. A precipitate is CaC₂O₄ (white).

TABLE V.

ANALYSIS OF GROUP V.

Testing for Magnesium, Sodium, Potassium and Ammonium.

Test portions of the *filtrate from Group IV.* for *magnesium, sodium* and *potassium.* Test a portion of the *original solution* for *ammonium.*

Magnesium.	Sodium.	Potassium.	Ammonium.
To a portion add am= monium hydroxide and chloride. Then Na_2HPO_4. A precip- itate is $MgNH_4PO_4$ (white).	Apply the flame test. Yellow obscured by the blue glass indi- cates sodium.	Apply the flame test. Violet not obscured by the blue glass in- dicates potassium.	To a portion of the original solution add excess of potassium hydroxide and boil. If ammonium compounds are present, they will be decomposed and ammonia gas, NH_3, lib- erated. This may be de- tected by its odor or by its action on moist red litmus paper.

TABLE VI.*

ANALYSIS OF GROUP III. (Phosphates present.)

Separation of Zinc, Aluminum, Iron, Manganese, Chromium, Nickel and Cobalt, also Barium, Strontium, Calcium and Magnesium.

To the *filtrate from Group II.* add *ammonium chloride*, then *ammonium hydroxide* till alkaline, and finally *ammonium sulphide.* Filter, wash and treat the *precipitate* as directed below.   *Take the filtrate to Table IV.*

Pour *hydrochloric acid* over the *precipitate* on the filter.   Examine any insoluble *black residue* for nickel and cobalt (see pages 44–46).   Divide the hydrochloric acid solution into two portions and treat each portion as shown below.

Portion I.

This may be examined for *iron, chromium, aluminum, zinc, manganese* and *magnesium.*   Add *chlorine water* and boil.   Divide the solution into two portions.

Portion A.	Portion B.
Test for **iron** as indicated in Table III.	To this portion add **ferric chloride** until a sample of it taken in a test tube gives a reddish precipitate of ferric hydroxide on adding ammonium hydroxide. Next evaporate the solution nearly to dryness; add water, then **potassium hydroxide** just **short of precipitation**, and finally an excess of **barium carbonate.** Let stand and filter.
Precipitate may contain **aluminum** and **chromium.** Boil the **precipitate** in an evaporating dish with **potassium hydroxide.** Filter.	**Filtrate** may contain **manganese, zinc, magnesium,** etc. Add **hydrochloric acid** and boil. Next add **ammonium hydroxide** to alkaline reaction, then **ammonium sulphide.** Heat and filter.

Precipitate.	Filtrate.
Test for **chromium** as indicated in Table III.	Test for **aluminum** as indicated in Table III.

Precipitate may contain **manganese** and **zinc.** Dissolve in a little **hydrochloric acid.** Add an excess of **potassium hydroxide** and boil. Filter.

Precipitate.	Filtrate.
Test for **manganese** as indicated in Table III.	Test for **zinc** as indicated in Table III.

Filtrate.

Test for **magnesium.** First add K_2SO_4, then NH_4OH and $(NH_4)_2C_2O_4$. Filter and test the filtrate for magnesium as indicated in Table V.

Portion II.

This may be examined for **barium, strontium** and **calcium.** Add **potassium sulphate** and filter.

Precipitate.	Filtrate.
Test for **strontium** by the Blowpipe Test (see page 47). Fuse the **precipitate** with Na_2CO_3 on platinum foil. Digest thoroughly with hot water. Filter. Dissolve the residue on the filter in acetic acid and add $K_2Cr_2O_7$. A precipitate is $BaCrO_4$ (**yellow**).	To the filtrate add **alcohol.** This precipitate may be $CaSO_4$. Filter, boil the precipitate with water, add NH_4OH and $(NH_4)_2C_2O_4$. A precipitate is CaC_2O_4 (**white**).

*This table is to be substituted for Table III. when phosphates are present in the solution. See note, page 70.

PART II.

Chemical Technology Applied to Dentistry.

CHAPTER X.

ALLOYS. GENERAL CONSIDERATION.

A chemical property of the metals, already referred
to in Chapter I., is that of uniting with one another
to form a class of compounds called *alloys*. As a rule,
this term refers to a combination of two or more metals effected by fusion. A class of alloys, however,
commonly called *amalgams*, in which mercury is one
of the constituents, may be produced not only by
adding mercury to a molten metal, but also by the
action of a metal on a salt of mercury, by the action
of mercury on a salt of a metal, and, finally, by bringing a metal in contact with mercury. Although the
term amalgam is applied to particular metallic compounds, it must not be inferred that they are distinct
from alloys. Indeed, they differ in no way except
that they contain mercury, which endows them with
certain peculiar properties. Hence, in the following
consideration, the general facts stated concerning the
chemistry of alloys and the classification quoted from
Matthiessen, apply to amalgams as well.

The knowledge of alloys, although perfected to a
high degree along certain lines, especially those which
relate to their superficial properties and to their
adaptability to the technical arts, must nevertheless
be considered very incomplete from the standpoint of

chemistry. Indeed, it is still a matter of controversy whether alloys are true chemical compounds. It is probable that metals in many instances combine with each other in definite proportions. It is a difficult matter in most cases, however, to determine the chemistry underlying these combinations, since the compounds, if formed, dissolve in an excess of the melted metals, and seldom can be separated in crystalline form. In a few instances alloys apparently possess definite chemical composition. Their formation is attended by phenomena which characterize chemical change, such as heat and incandescence. Metallic crystals are at times formed, and in nearly all cases there is observed a variation in specific gravity and in melting point from the corresponding values of a mere mixture of the metals. Furthermore, there are some instances of alloys found in nature showing the constituents in definite proportions. Thus an amalgam of silver occurs in which the metals are combined in the ratio of their combining weights, and hence can be represented by the formula HgAg.

Although other evidence could be cited which would seem to support the theory of chemical union, it nevertheless must be admitted that whenever combination takes place between metals the alteration of physical properties, which is a distinctive characteristic of chemical union, is not very pronounced. In the most unquestionably metallic compounds the properties which characterize metals, such as color, luster, hardness, conductivity, etc., although often modified, are still retained; and in alloying no such loss of identity takes place as occurs when a metal unites with a nonmetallic element, such as oxygen, chlorine, etc.   Moreover, the affinity which binds the metals in an alloy is very feeble, and the state of combination is, therefore, easily affected by outside forces.   Many alloys, including the most stable amalgams, are readily decomposed by heat, and some simply by air or by water.   Not a few alloys separate into layers, especially if allowed to cool slowly, with the result that an ingot not at all homogeneous is formed.   This phenomenon, which is a matter of great importance in refining certain metals and in preparing alloys for industrial purposes, is commonly called *liquation*.   A good example is the separation of lead and zinc, two metals which apparently have little affinity for each other.   If equal parts of these metals are fused and then thoroughly mixed and allowed to cool slowly in a deep mold they will be found to separate almost completely, only one and six-tenths per cent of zinc being retained by the lead.   An important application of liquation in the refining of metals is shown in the separation of silver from lead by the Pattison method described on page 20.   The author has had occasion to make some experiments with alloys of silver and tin, such as are commonly employed with mercury in filling teeth, and he finds that by keeping a perfectly homogeneous alloy of these metals in a molten state for one-half hour in a deep graphite mold a decided liquation results.   This is shown by the fact that a large proportion of the silver employed is found at the bottom of the ingot. In cases where liquation has occurred a more homogeneous alloy can be obtained by remelting.

According to Matthiessen, it is probable that an alloy of two metals in a melted state is, first, a solution of one metal in another; or second, a chemical combination; or third, a mechanical mixture; or fourth, a solution or a mixture of two or of all of the above.

Physical Properties of Alloys.

From the preceding study of the metals it may be observed that comparatively few possess properties which render them suitable to be employed in the pure condition in the arts and manufactures. By properly alloying a metal, however, with one or more others it is possible to obtain a new metal, as it were, which more nearly possesses the required properties than do the metals entering into its composition. Thus, attention has been called to the fact that gold and silver are too soft to be minted or to be used in jewelry, but that the addition of certain quantities of copper renders them harder and more capable of resisting wear. Copper, which in the pure state is too tough to be worked in the lathe, is converted when alloyed with zinc into the useful product known as *brass*, the properties of which make it particularly suitable for turning. Moreover, many metals, as arsenic, antimony, bismuth, manganese, chromium, etc., which alone are practically without value, confer certain desirable properties upon alloys. The changes in the properties of metals induced by alloying have been quite fully discussed in Chapters I. and II., but for the convenience of the student they are briefly reviewed here.

COLOR, LUSTER AND SONOROUSNESS.

Those metals which possess decided color, namely, copper and gold, are greatly modified by alloying, and often a colored alloy is obtained by combining two metals which possess no decided color. When ten per cent of aluminum is added to copper an alloy resembling gold in color is produced, and when antimony and copper in equal proportions are fused together a violet alloy is formed.

The luster of a metal is often modified, and at times even destroyed by alloying. The case of gold alloyed with base metals is an example.

A property possessed to only a slight degree by most metals, namely, *sonorousness*, is greatly increased in many instances by alloying. Thus copper and tin combined in certain proportions produce the alloy known as *bell metal*. Copper and aluminum also produce a very sonorous alloy.

SPECIFIC GRAVITY.

Contrary to what might be expected, alloys seldom possess a specific gravity which corresponds to the mean of the specific gravities of the constituents. In cases where a perfectly homogeneous alloy has not been obtained, as in liquation, the specific gravity will vary, of course, in different portions of the ingot.

FUSIBILITY AND CRYSTALLINE FORM.

In most cases the melting point of an alloy is less than the mean melting point of its constituents. Indeed, it is not uncommon to find the melting point greatly below that of any of the metals employed in the alloy. Common soft solder, for example, melts more easily than either the tin or lead composing it. The compounds known as *fusible alloys* owe their great fusibility to the presence of bismuth, cadmium or mercury. Thus an alloy known as Rose's metal, composed of one part of lead, one part of tin and two parts of bismuth, melts at 95° C., or considerably below the temperature of boiling water. Another alloy, composed of eight parts of lead, fifteen parts of bismuth, four parts of tin and three parts of cadmium, melts at 65° C., the theoretical point shown by calculation being 284° C. An alloy of the alkali metals, sodium and potassium, is liquid at ordinary temperature.

Alloys often exhibit a crystalline form. Like metals, they become crystalline under the influence of percussion and other forms of mechanical working.

MALLEABILITY AND DUCTILITY.

These properties are almost always diminished and in many cases destroyed.. Even the union of two very malleable metals, as gold and lead, forms an alloy which is very brittle. Again, the combination of a brittle and a ductile metal produces an alloy which is low in ductility. Examples are the alloys of gold with antimony and bismuth, already referred to in Chapter II.

HARDNESS, ELASTICITY AND TENACITY.

The effect of alloying upon hardness has been already referred to. The alloy of gold and silver used for coin is a good example.

The elasticity of certain metals may be increased by the addition of small quantities of other metals. An example is the alloying of gold with platinum.

As a rule, tenacity is greatly increased by alloying. Thus the tenacity of copper is increased threefold by adding twelve per cent of tin.* The tenacity of iron compared with that of steel is in the ratio of one to two and one-half.*

CHANGE OF VOLUME WITH TEMPERATURE.

The coefficient of expansion of alloys with heat is approximately the average of the constituent values. However, a few decided variations from this rule have been observed. Thus a copper-tin alloy expands less than pure copper, although the expansion of tin is considerably more than that of copper.

SPECIFIC HEAT AND CONDUCTING POWER.

The specific heat of an alloy is the mean of the specific heats of the metals composing it.

In their power of conducting heat and electricity alloys do not in any case exceed the conductivity of the components. In fact, alloying generally results in reducing these properties.

Chemical Properties of Alloys.

The chemical properties of alloys do not differ materially from those of the metals constituting them.

---

*Results obtained by Matthiessen.

In certain instances the tendency to oxidize is increased by alloying. Alloys of lead and tin are oxidized more readily than either of the constituents. Alloys of silver and tin, those used with mercury for filling teeth, are exceedingly oxidizable, as will be shown later. Probably the most important and interesting change in the chemical properties of certain metals after alloying is that shown by their action toward solvents. Silver, which by itself is readily soluble in nitric acid, becomes insoluble when alloyed with much gold. Indeed, as already stated, it is difficult to remove the silver completely from gold unless the latter constitutes less than twenty-five per cent of the alloy. Again, platinum, which by itself is insoluble in nitric acid, dissolves completely in this acid when alloyed with ten or twelve times its weight of silver.

Preparation of Alloys.

The method commonly employed in preparing alloys consists in fusing the proper proportions of the metals in a crucible. Usually graphite crucibles are used instead of those made of clay or sand, as they are less liable to crack when heated to a high temperature. The preparation of alloys is not, in most cases, a difficult matter, especially if certain important properties of the metals, such as fusibility, specific gravity, tendency to oxidize, etc., are kept in mind. Often the constituent metals are mixed and fused, particularly if they do not differ widely in their melting points. In most cases, however, it is better to melt the metal possessing the highest melting point and then add the others in the order of their fusibility. In order to prevent oxidation the surface of the metals should be covered with some substance, as borax or powdered charcoal; the heat should not be continued for any great length of time after the alloy is in a molten state; and finally, the temperature of the furnace should always be adapted to the fusibility of the metals, i. e., a heat sufficient to melt gold or silver should not be employed in making an alloy of tin and lead. In all cases care should be taken to prevent liquation, by stirring the alloy thoroughly with a pine stick and pouring into a cold mold. More details concerning the preparation of certain alloys will be given later. Although the directions given above are applicable in the preparation of alloys in general, it should be borne in mind that no fixed rule can be followed in all cases.

CHAPTER XI.

APPARATUS.

Various forms of apparatus, some of which are not commonly included in the ordinary laboratory outfit, are essential to the work outlined in subsequent chapters. In order that the student may become somewhat familiar with them at the outset, considerable space will be devoted to their description.

I. Balances and Weights.

Analytical Balance. (Fig. 2.) This is an extremely delicate balance used in accurate chemical and physical work. The various parts are an upright pillar, on the top of which is a highly polished steel or agate plane. Resting on this plane and attached to the center of the beam is a V-shaped steel or agate knife edge about which the beam vibrates. At each end of the beam is attached another similar knife edge, in this case inverted (Λ), supporting a little stirrup with a steel or agate plane, and from this stirrup is suspended the light wire frame carrying the scale pan. This form of rest and support is designed to minimize the friction and to increase the sensitiveness of the instrument. In order to prevent the knife edges from being injured by constant wear or jar on the planes, the beam is raised and supported when not in use by a lever mechanism shown below the beam in the illustration. This is operated by a milled

FIG. 2.

handle in front. In addition to the beam support almost all balances have arms (see illustration) called "bumpers" which arrest the scale pans from beneath while the beam is resting upon its bearings. These are manipulated by an ivory push-button to the left of the milled handle.

At the base of the pillar is an ivory scale in front of which swings the point of the index needle attached to the center of the beam. At each end of the beam is a small adjusting screw by means of which the beam is equipoised. The right side of the beam is usually divided into ten large and one hundred small scale divisions. By sliding a small piece of platinum wire, called a "rider," along the graduated portion the final adjustment of equilibrium is made while weighing and the use of inconveniently small weights is avoided. With a ten milligram rider each large division represents one milligram (0.001 gram), while a small division represents 0.0001 gram. When the rider rests on the tenth large division a ten milligram weight placed in the left-hand scale pan should be exactly counterpoised. The rider is manipulated by the sliding rod shown above the beam and extending outside the case.

Weights. The metric system of weights (see Appendix, Section I.) is used. The weights themselves are enclosed in a wooden case (Fig. 3) and usually range from 100 grams to 10 milligrams. Weighings below ten milligrams are obtained by means of the rider. The gram weights are commonly made of brass and the fractions of a gram, of platinum. After using a weight always return it to its proper place in the case but never employ the fingers in doing so. *In handling weights use a pair of tweezers.*

Special Directions. It should be borne in mind that the balance is a very delicate instrument and to give good results it must be used with the utmost

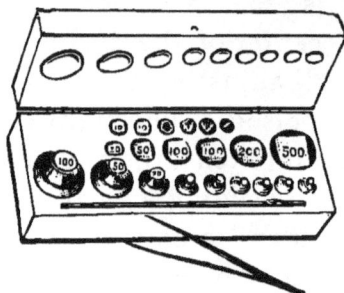

Fig. 3.

care. In weighing take a position directly in front of the index needle, raise the sliding window carefully and place the substance to be weighed in the left-hand scale pan. Estimate the weights required and place them in the right-hand pan while the beam is still supported. Carefully lower the support and leave the pans free to swing in order to see which carries the greater weight. Add or remove weights as necessary, always supporting the beam while doing so. When the smallest weights fail to counterpoise the substance employ the rider. Lower the window to avoid drafts of air, which might interfere with the accuracy of the weighing and cause the index needle to vibrate; perfect equilibrium is reached when it goes to an equal distance on each side of the central mark on the ivory scale. Support the beam, remove the weights and record the result. Many of the details involved in weighing cannot be described, and can be learned only by experience.

When the balance is not in use keep the case

closed, and never leave the beam unsupported. Remove dust from the pans and other parts with a fine brush. Never bring corrosive substances in con-

FIG. 4.

tact with the pans; use a counterpoised watch crystal. Keep a beaker or other vessel containing dry calcium chloride inside the case to absorb moisture. Protect the balance from direct sunlight, from acid fumes and from the vibrations of the building. When not in a level position, which will be indicated by the spirit-level at the base of the upright pillar, adjust by means of the leveling screws underneath the case. The equipoise of the beam can be adjusted by the screws at each end.   Study the balance and become familiar with its different parts and its manipulation.   A good analytical balance will carry 100 grams in each pan and is sensible to one-tenth of a milligram.

Pulp Balance.   (Fig. 4.)   This is a style of balance adapted to weighing fluxes, metals for alloys, and to other uses in which considerable accuracy is required.

It will be observed that the balance illustrated does not differ greatly in its general construction from the analytical balance; it is provided with knife edge bearings, movable pans, set screws and level, and an eccentric for raising the beam.   Such a balance is capable of carrying 300 grams in each scale pan, and of turning with a five milligram weight. Keep the pans free from dust and do not bring corrosive substances in contact with them.

FIG. 5.

Laboratory Scales.   (Fig. 5.)   These are scales for coarse weighing.   Those illustrated (Harvard Trip) have porcelain plates in place of metal scale pans. Their capacity is 1000 grams (one kilogram).   Weighings from five grams to one-tenth of a gram can be made by using simply the beam and sliding weight in front.

II. Furnaces* and Accessories.

Hoskins' Assay Furnaces. Very satisfactory furnaces for assaying purposes and for use in refining and alloying on a comparatively large scale are those known as Hoskins' Hydro-carbon Assay Furnaces. The outfit, shown complete in the Frontispiece (Fig. 1), consists of a crucible furnace, a muffle furnace and a patent blowpipe, employing gasoline as fuel. The different parts are illustrated and described below.

FIG. 6.

Crucible Furnace. (Fig. 6.) This is a furnace in which crucibles are employed. Various sizes of Hoskins' crucible furnace are made. Size No. 1 is cylindrical

* For descriptions of various furnaces not considered here refer to Essig's "American Text-book of Prosthetic Dentistry," or to the catalogues of Richards & Co., Chicago, E. H. Sargent & Co., Chicago, and of other dealers in chemists' and assayers' supplies.

in form, and contains a chamber six inches high and four inches in diameter. A convenient size (No. 4) and form for a large laboratory is that illustrated, containing a chamber six and one-fourth inches high, eight inches long and six and one-half inches wide, and heated by a No. 3 blowpipe, holding one gallon of gasoline. The furnace body is made of an infusible material, enclosed in a casing of sheet iron. At the left is shown the fire-hole or opening for the blast, and on top the chimney for the escape of the products

Fig. 7.

of combustion. The heat can be made to range from that of the Bunsen burner to that required to melt cast iron, and the maximum effect can be produced in ten minutes from the start. The advantages of Hoskins' furnaces over those employing solid or gaseous fuel are many; dust, ashes, smoke and radiated heat, connected with the use of coke and coal, are avoided, and a high degree of heat can be produced in a very short time; unlike furnaces burning ordinary illuminating gas, the blast is automatic, not requiring the use of a blower. Finally, they can be used in places where gas is not obtainable.

Muffle Furnace. (Fig. 7.) This is a furnace used in the processes in assaying known as cupellation and scorification, and in other operations not requiring the heat of a direct flame, as enameling, etc. It consists of a furnace body in which is an arched clay enclosure known as a muffle. The Hoskins' furnace illustrated is heated from the rear by the blowpipe used in connection with the crucible furnace. A convenient muffle furnace is one taking a muffle ten inches long, six inches wide and four inches high.

FIG. 8.

Hydro-carbon Blowpipe. (Fig. 8.) The important parts of this apparatus are : A tank, T, which contains gasoline and compressed air, the latter being forced in by the pump, P; an automatic valve at A, which closes after each stroke of the pump and a cut-off valve, C, which closes completely the connection between the pump and tank; a plug, F, removed when the tank is to be filled; a vent, V, for releasing the air pressure in the tank when the apparatus is not in use; a pipe, H, connecting the tank and the burner, D; a regulating valve, E, terminating in a fine point controlling the flow of gasoline; and two boxes, SS, packed to prevent leaking about the valves.

To operate the blowpipe, unscrew F and fill the tank about two-thirds full of gasoline (74° Beaumé). *Caution:* Do not attempt to do this near the highly heated furnace or near any flame. Replace F, open C, give the pump six or eight strokes and close C. Bring the nose of the burner within two inches of the

FIG. 9.

fire hole in the furnace, and place a Bunsen flame under D. After ten minutes or so the gasoline in the burner tubes will become vaporized, and if E is slowly opened the gas will escape and can be ignited. Regulate the intensity of the flame by the pump and valve E. When the interior of the furnace has reached a bright red heat, quickly shut off E and immediately turn on again. When the furnace is hot enough the escaping gas ignites and burns inside the furnace. If, however, the furnace is not hot enough the gas will fail to ignite. In such a case continue the *first* flame until the proper temperature is reached. *There is no danger connected with the use of this blowpipe except through gross carelessness.*

Other Furnaces. Although an outfit like that just described is indispensable in many operations, a smaller furnace can be used to advantage in certain cases, particularly in alloying small quantities of metals, in calcining cement powders, etc., etc. Indeed, the furnaces illustrated and described below will meet most requirements of the student and the practicing dentist. A very efficient and convenient furnace (Fig. 9) for melting 300 grams or less of metal is that made for dentists and jewelers by the Buffalo Dental Manufacturing Company. The furnace body is composed of a substance much lighter in weight than fire clay and said to possess only one-tenth its conducting power for heat. As illustrated, the apparatus consists of a furnace body with a cover and a blowpipe employing gas, all mounted on a cast iron base. With a gas supply pipe of three-eighths of an

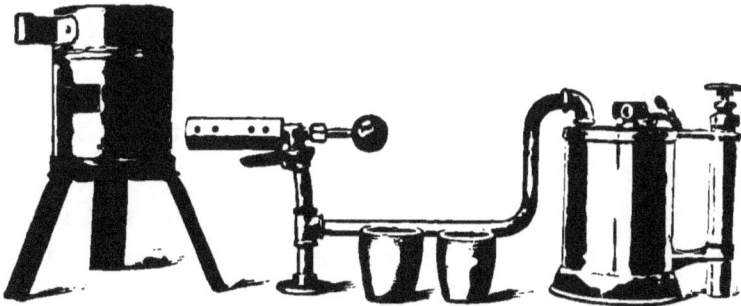

FIG. 10.

inch and a good foot-blower (Fig. 24) sufficient heat can be obtained in ten or twelve minutes to melt cast iron.

A furnace (Fig. 10) fully as efficient as the one just described and much more convenient in that it does away with the foot-blower, is that arranged by the author for the use of students in his laboratories. As will be observed, the furnace body is that illustrated in Fig. 9, mounted on a suitable tripod. The blow-pipe is a common gasoline paint burner with the tank removed some distance from the burner (the burner ordinarily is attached directly to the tank) in order to avoid heat from the furnace. To operate the blowpipe, unscrew the plug on top, fill the tank about two-thirds full of gasoline and replace the plug. Open the valve in the burner until the cup below is full of gasoline, then close; ignite the gasoline in the cup and when it is nearly burned, open the valve again. A strong, steady blast flame results. With a single filling, one quart of gasoline, this blowpipe will burn for two hours and during that time will require pumping up but once or twice. Some kerosene burners are now made which give about as hot a flame as those burning gasoline. These may be used unmodified, as it is not necessary to separate the tank from the burner.

Crucibles and Crucible Tongs. Crucibles (Fig. 11) in which metals are fused are made of graphite (black lead or plumbago), French clay, Hessian sand, etc. The so-called clay or sand crucibles are made in round and triangular forms and are suited to nearly all metallurgical operations. They

Fig. 11. seldom can be used more than two or three times, as they crack when heated. Graphite crucibles last indefinitely and can be used in all

fusions in which no oxidizing agents, as potassium nitrate, are used. A crucible for the small furnaces should be about two inches high and two inches

Fig. 12.

in diameter. The Hoskins' furnace takes four crucibles five and one-fourth inches high and three inches in diameter. Tongs for handling crucibles are illustrated in Fig. 12.

Scorifiers and Scorifier Tongs. Scorifiers (Fig. 13)

are shallow clay dishes used for scorifications. The size ordinarily used is two and

Fig. 13. three-fourths inches in diameter. A convenient form of scorifier tongs is that shown in Fig.

Fig. 14.

14. The forked arm fits the bottom of the scorifier and the straight arm extends across the top.

Cupels and Cupel Tongs. Cupels (Fig. 15) are small

articles made of bone ash and used in the process of cupellation. Bone ash absorbs the oxides of almost all metals, particularly those of lead; hence it is used in purifying

Fig. 15. gold and silver, which do not oxidize. A good cupel will absorb nearly its own weight of lead oxide. A convenient size for assaying purposes is one

and a half inches in diameter.   Cupel tongs are
illustrated in Fig. 16.

FIG. 16.

Button Mold.   (Fig. 17.)   The button mold, often
called a slag or scorification mold, is generally used

FIG. 17.

in pouring scorification or other fusions in which a
button-shaped ingot is desired.   This mold should
be kept clean and should be warmed on the furnace
before using.

Bullion Mold.   (Fig. 18.)   This is a convenient

FIG. 18.

mold for casting a bar-shaped
ingot of gold or silver bullion,
solder, amalgam-alloy, etc.
It is provided with a sliding
partition, so that any length below eight inches can
be obtained.   For many purposes a mold made of
soapstone can be substituted for the one just described.
When alloys are to be poured the mold should not
be warmed, as an alloy should solidify quickly after
pouring in order to prevent the metals from separating.

Upright Ingot Mold. (Fig. 19.) This gives an ingot about one-eighth of an inch in thickness, which can be placed in the rolling mill without hammering out. The right-hand portion of this mold, as illustrated, is adjustable so that ingots of different widths can be obtained.

FIG. 19.

III. Miscellaneous Apparatus.

FIG. 20.

Hand Rolling Mill. (Fig. 20.) When an ingot is to be rolled or laminated it is first annealed and

then passed between polished steel rollers, so controlled by pressure screws that they can be brought
closer together each time the ingot is passed through.
The thickness of the resulting plate is determined by a

Fig. 21.

gauge (Fig. 21).   During the rolling the ingot should
be annealed several times to prevent it from splitting
at the edges.   When a mill without geared pressure

Fig. 22.

screws is employed care
should be taken so to regulate the rollers that a plate
of the same thickness on
both sides is produced, else
it will curve.   However,
should this happen, screw the rollers closer together
on the thickest side, and continue rolling as before.
Do not reverse the plate, for this will crumple it.

Iron Retort. (Fig. 22.) This is used in distilling mercury from amalgams. A good substitute for this apparatus can be made at any plumber's shop.

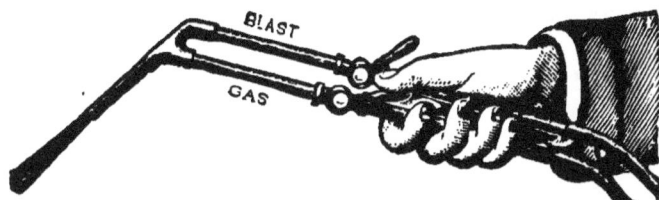

FIG. 23.

Blowpipe and Blower. In many operations, particularly in melting small quantities of gold, silver, etc., a blowpipe giving a higher temperature than the ordinary mouth blowpipe is required. For this purpose a brazing blowpipe (Fig. 23) can be used and the blast required can be obtained by a foot-blower (Fig. 24).

FIG. 24.

In operating attach the tube marked "gas" to the gas supply pipe by means of rubber tubing, and in the same manner connect the tube marked "blast" to the blower. By manipulating the valves as shown, both air and gas can be regulated and the character of the flame can be controlled.

Hydrogen Generator. (Fig. 25.) This is also known as Marsh's apparatus. When used in testing

FIG. 25.

for arsenic, antimony and tin, as indicated in Chapter V., a straight piece of glass tubing is attached to the exit tube by means of rubber tubing, as illustrated. In case Marsh's test for arsenic or antimony is to be made, a tube, drawn out until the opening in one end is about the size of the lead in a pencil (see figure), is substituted for the straight tube.

Porcelain Crucibles and Pipestem Triangles. Fig.

FIG. 26.

26 represents a porcelain crucible or capsule, with cover, used for many purposes in quantitative chemistry. They are designed to withstand high heat. A convenient size is one and one-half inches in diameter. Fig. 27 represents a triangle made of wire covered with pipestem. This is used to support the porcelain crucible while being heated.

Other utensils so common as to require no descrip-

FIG. 27.

tion are : Anvil and vise, separate or attached; hammers, a light one, for hammering out gold and silver beads, and a heavy one for general use; pinchers with pointed nose for handling gold and silver beads; shears for cutting metal; files, fourteen-inch bastard, for filing alloys; steel brush for cleaning files; magnet for removing iron filings from amalgam-alloys; sieves, forty, sixty, eighty meshes, for sifting filings; bolting cloth for sifting cement powders; mortars, a small wedgwood mortar for mixing amalgams, a large mortar for grinding purposes; amalgam-filling instruments, spatula and glass plate for mixing cement, etc.

IV. Apparatus for Testing Amalgams.

Micrometer. The apparatus illustrated in Fig. 28 has been designed by the author for the use

FIG. 28.

of students, in determining the *expansion* and *contraction* of amalgams employed in filling teeth. It con-

sists of a heavy base with two upright posts; between
the latter is a horizontal axis with cup and pivot

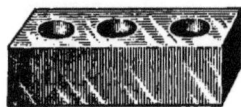

bearings, and attached to the center
of this axis is a long steel needle
which vibrates in front of a scale.
Near the axis and attached to the
needle is a vertical plunger held in

FIG. 29.

position by a guide extending out from one of the
posts. It is evident that any vertical movement of this
plunger will be greatly magnified at the point of the
needle.   The amalgam to be tested is packed in
cavities in a steel block (Fig. 29), and the latter is
placed in a guide attached to the base; the plunger
is allowed to rest upon the surface of the amalgam
and the point on the scale indicated by the needle is
recorded.   At various intervals during several days
the block is returned to the apparatus and any expansion or contraction of the amalgam filling is noted by
the movement of the needle along the scale.   It is
obvious that, owing to irregularities on the surface of
the filling, the plunger should rest on the same point
in every measurement taken.   To accomplish this, a
mark, corresponding to one on the guide, is made on
the steel block when the first measurement is taken.
At the left is seen a weight; this nearly counter-
poises the needle and thus prevents the plunger from
sinking into the amalgam while still soft.   The scale
is graduated in such a manner that one small division
indicates $\frac{1}{60}$ of a millimeter (approximately $\frac{1}{1500}$ inch)
at the plunger; and as it is an easy matter to read
to one-quarter of a scale division, an expansion or

contraction of $\frac{1}{200}$ millimeter (approximately $\frac{1}{5000}$ inch) can be accurately measured.

Dynamometer. Fig. 30 represents an apparatus which may be used in determining the strength of amalgams and the change of form which they undergo when subjected to pressure, i. e., their so-called "flow." The apparatus consists of a steel spring, CC, in the form of a double bow. When this is compressed by the screw shown at the right a pressure

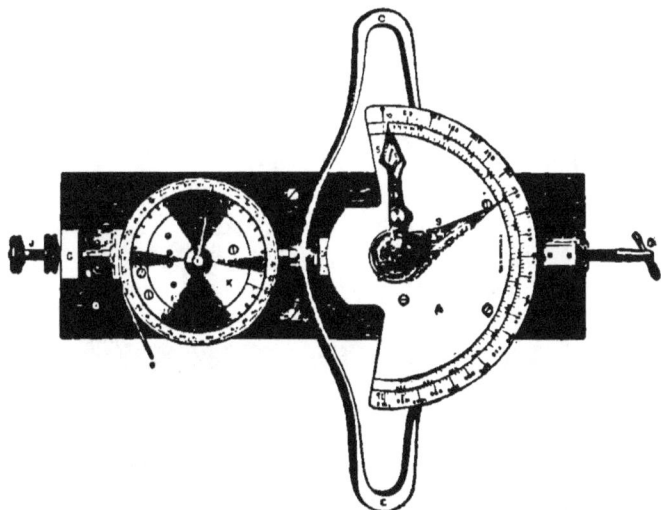

FIG. 30.

is communicated by a steel rod to the amalgam, which in the form of a block is placed between the end of the rod and the point of the stationary screw, J. By a cog mechanism attached to the vertical needle, S, the pressure exerted is recorded in pounds avoirdupois on the inner scale of the large dial, A. (The outer scale cannot be used in these tests, as it is designed for another purpose.) The vertical needle, S, in passing about the scale carries with it the second needle, also marked S. When the block of amalgam crushes the first needle is carried back by a coil spring to the zero mark, while the second, not being connected with the pinion, remains to indicate the stress applied.

The amalgam to be tested is packed in the small cavities (lettered A, B, C, D) of the matrix (Fig. 31). When these blocks of amalgam, which have the

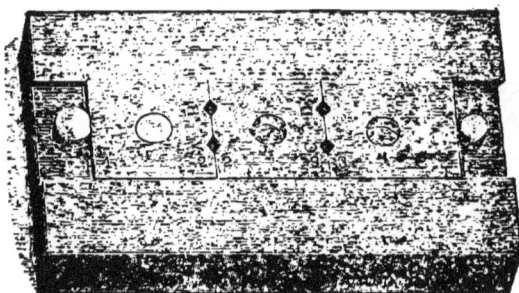

Fig. 31.

dimensions 0.085 x 0.085 x 0.085 inch, have "set" they are removed by withdrawing one of the pins at the end and removing the sliding sections. After five days, tests may be made upon these samples. A block which is to be tested for strength is placed in the instrument and the screw turned until the sample crushes. The pressure necessary to do this will be recorded as already described, and this result may be taken as representing the crushing-strength of the amalgam. A block which is to be tested for flow is placed in the instrument in the same manner as above and a pressure equal to one quarter of that required to crush the first sample is applied for some time. The flow of the amalgam is indicated in thousandths of an inch by the needle in the small dial, K. In all cases the time and pressure employed in bringing about a certain flow should be stated, since both these quantities are factors in the result. Finally, the flow is best stated in percentage. For example, if a block which originally was 0.085 inch in thickness is reduced to 0.0765 inch by applying a pressure of fifty pounds for one hour, then the complete statement of the result would be: Pressure, fifty pounds; time, one hour; flow, ten per cent.

V. Measuring Apparatus.

Cylinders. (Fig. 32.) These are employed in measuring liquids when great accuracy is not required. They are made to contain 1000 c. c. (cubic centimeters), 500 c. c., 250 c. c., etc. Each cubic centimeter is divided into tenths.

Flasks. (Fig. 33.) These are used in place of cylinders in measuring liquids when great accuracy is required. They are made to contain

FIG. 33.

FIG. 32. 1000 c. c., 500 c. c., 250 c. c., etc., of liquid at a certain temperature, usually 15° C.

Pipettes. (Fig. 34.) These are used in measuring accurately a certain volume of liquid. The

liquid is drawn up by suc- tion to an indicating mark on the tube above the bulb and retained by closing the opening of the tube with the finger. The pipette illustrated delivers 50 c. c. Others are made to deliver 100 c. c., 25 c. c., 10 c. c., 5 c. c. and 1 c. c. Sometimes they are made after the form of a burette.

Burettes. (Fig. 35.) These are made to deliver accurately volumes of liquid from 100 c. c. down to one-tenth or one-twentieth of a cubic cen- timeter. Some are provided with a ground glass stopcock, others with a rubber tube, compressed by means of a clamp. The illustration shows a burette filled with a liquid, and mounted on an iron stand, ready for use. Burettes and other glassware, when very dirty, can be cleaned with a mixture of potassium dichro- mate solution and sulphuric acid, after which they should be rinsed thoroughly with distilled water.

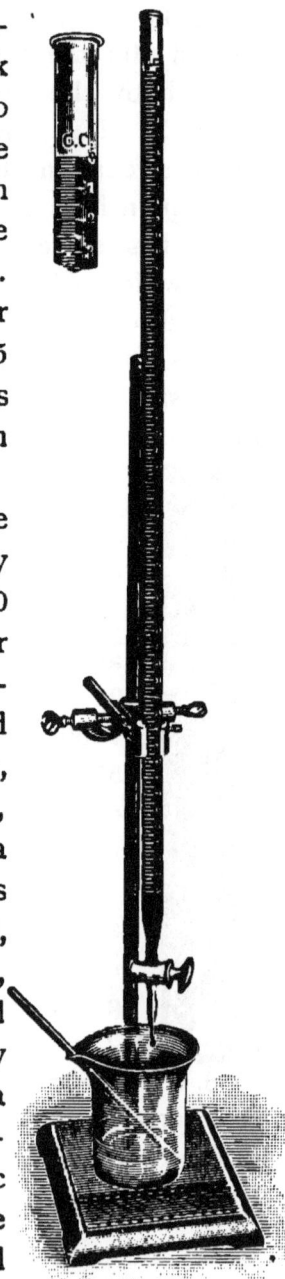

Fig. 34.

Fig. 35.

VI.   Urine Analysis Apparatus.

Urinometer. (Fig. 36.) This is an instrument employed in determining the specific gravity of urine. It floats in the urine, being held in an upright position by a bulb of mercury at the lower extremity. The degree on the graduated stem which corresponds with the surface of the urine is taken as the specific gravity. As a urinometer is graduated to give the correct reading only at a definite temperature, care should be taken to bring the urine to the required temperature before immersing the urinometer in it. This temperature, usually 15.5° C. (60° F.) or 25° C. (77° F.), is generally marked on the graduated stem of the instrument.

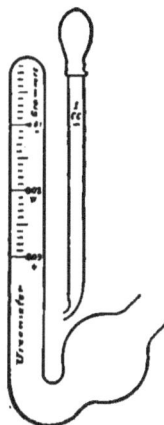

FIG. 36.

FIG. 37.

Doremus' Ureometer. (Fig. 37.) This is an apparatus used in determining the quantity of urea in urine. The apparatus is filled with a hypobromite solution and

1 c. c. of urine is introduced into the upright portion by means of the little pipette. The urea is decomposed by the hypobromite and the resulting nitrogen accumulates in the graduated portion of the apparatus where the fraction of a gram of urea in 1 c. c. of urine is read off. The urine should always be introduced slowly in order to prevent a loss of nitrogen through the opening of the bulb.

CHAPTER XII.

REFINING GOLD, SILVER AND MERCURY.*

The refining of metals presents a problem of vast industrial importance, and one which has considerable practical bearing upon dentistry. In order to illustrate the chemistry underlying this art, and at the same time to provide pure metals for subsequent use in making alloys, simple methods are here given for the student to follow in refining different forms of scrap gold and silver and in removing various impurities from mercury. At least two grams of gold and twenty-five grams each of silver and mercury should be refined.†

Refining Gold.

Gold accumulated by the dentist and jeweler may be classified as follows:

Class I. Clean scrap, as plate scrap, clippings, filings, etc. Waste of this character seldom needs refining—simply remelt, raising or lowering the carat if not of suitable fineness originally (see Appendix,

---

*Material for this work, such as discarded gold jewelry and other scrap gold, silver watch cases, spoons, mutilated coin, waste dental amalgam, crude mercury, etc., can be obtained of refiners.

†The form of report to be submitted upon the completion of this work is shown in the Appendix, Section II.

Section I.) and roll into plate.   Follow the instructions given in Method I.   Should it be desired, however, to reduce the scrap to pure gold follow Method III.

Class II. Mixed scrap, as clippings, filings, solder, base metals, etc., containing little or no platinum. This scrap should be freed as far as possible of foreign substances, then melted down as indicated in Method I., to remove base metals, and finally refined by Method III., in order to remove silver.   In case the scrap contains much platinum, remove the latter mechanically, as far as possible, and then follow the instructions given in Method II.

. *Class III. Sweepings,* as bench sweepings, residues left after incinerating sandpaper strips or disks used in polishing gold, etc., etc.   Waste of this character will usually contain considerable foreign matter, as plaster of Paris, porcelain, base metal, amalgam, etc. These substances should be removed and the waste then reduced to ashes, using no fluxes. Mix the residual matter with about one-quarter its weight of saltpeter (powdered). Place the mixture in a red hot clay crucible (the crucible should not be more than half full), and cover with a thin layer of salt to. prevent frothing.   Keep the crucible at a very high heat until the slag is thin enough to allow the metal to settle. This is determined by plunging an iron rod into the crucible. When removed the slag should run freely from the rod. In case, however, the slag refuses to become thin, add more saltpeter and a little borax, and increase the heat if possible.   Finally, remove the crucible from the furnace, tap it gently on some solid object, let it cool undisturbed, then break it open and remove the metal, which should be collected in the form of a button in the bottom, if the work has been conducted properly. In case the metal is bright and malleable it may be treated at once by Method III. If, however, the button is dull and brittle, it still contains some base metal. Hammer or roll out the button, cut in small pieces, place in a clean clay crucible and "saltpeter," as indicated in Method I., until clean. Finally, refine by Method III.

METHOD I.

Select a clean clay crucible and heat it in the furnace until red hot. Introduce the fragments of·gold to be remelted, and add a small quantity of saltpeter (crystals). Replace the furnace cover, and so regulate the heat that the gold will melt as quickly as possible. In the meantime clean the interior of the upright ingot mold (Fig. 19) with a cloth, and then oil it. Common machine oil will answer the purpose, but do not use inflammable oils. Place the mold on top of the furnace to warm. The oil protects the gold from contamination with iron and the heat prevents the molten metal from cooling too rapidly and thus forming an ill-shaped ingot. When the gold is melted and ready to pour, place the mold in a convenient position, remove the crucible by means of the crucible tongs (Fig. 12), and pour the metal not too rapidly into the mold. Immediately remove the ingot and wash it, using for this purpose soap and a stiff scrub-brush. Next anneal by holding it in a flame, or by placing it in the furnace. When a cherry red color is reached plunge it into a dilute solution of

sulphuric acid and again wash, dry, and proceed to
roll it out (see page 111), reannealing at times during
the operation.   Should it happen that the ingot is brittle and refuses to roll without cracking cut it in small
pieces, place the pieces in a clay crucible and subject
to the highest temperature of the furnace.   When
molten, raise the crucible to the top of the furnace by
means of the tongs; throw small crystals of saltpeter
into the metal, and give the crucible a gentle circular
motion to insure thorough mixing.   Continue the heat
for some time, and finally again pour into the upright
mold.   If the ingot still persists in cracking, repeat
the operation just described.

METHOD II.*

When the larger scraps have been cut in small
pieces, place them together with the filings, etc., in a
flat bottom flask, designed to withstand heat, and
cover with aqua regia.   In making the acid use two
parts of concentrated hydrochloric to one part of concentrated nitric acid.   Place the flask on a sand bath,
apply heat gently and at times add small portions of
acid to maintain the reaction.   It may happen that
some fragments of gold become coated with a film
of silver chloride and refuse to be acted upon
by the acid.   In such a case decant the acid liquid,
rinse the flask with a little water and boil the
fragments with ammonium hydroxide until the

*This method, if carefully followed, will yield practically
chemically pure gold.

film of silver chloride is removed. Carefully decant the liquid, replace the acid and continue applying the heat until the gold is completely dissolved. Silver present in the original substance remains as a residue in the bottom of the flask. Remove this by filtration, but before doing so add enough water to weaken the acid somewhat. Wash the last trace of the yellow gold solution from the filter paper. The residue on the filter is silver chloride and may be reduced to silver by fusing on charcoal with sodium carbonate. The gold chloride solution should now be transferred to a large beaker or jar and diluted with water until only faintly acid in taste. Prepare a clear solution of ferrous sulphate, roughly calculating that five parts of the sulphate crystals are required to precipitate one part of gold. Add this slowly to the gold solution and allow the resulting brown precipitate of metallic gold several hours in which to settle. When the supernatant liquid has become clear add a little more ferrous sulphate to insure complete precipitation. Decant the clear liquid and boil the gold with dilute hydrochloric acid to remove iron. Pour the acid on a filter and continue boiling with fresh portions of acid until satisfied that all traces of iron are removed. Finally, transfer the gold to the filter and wash with hot water until no acid is detected in the washings. Dry in an air bath or in a porcelain dish over a burner; incinerate the paper and fuse the metal with the addition of a little borax and saltpeter on charcoal with a blowpipe (Fig. 23), or in a clay crucible in the fur-

nace.   Gold, carefully prepared as indicated, is pure
and should be very soft and malleable.   Should it
be found brittle, however, likely due to imperfect
washing, "saltpeter" it as indicated in Method I.

METHOD III.

Method I. is designed to remove base metals only.
Although Method II. separates every metal from the
gold, it is a tedious operation and need seldom be
employed except in cases where platinum is present in
the scrap in considerable quantity.   The method
about to be described, commonly called *quartation*,
is very convenient and rapid and is applicable in
cases where silver, copper, and indeed small quantities of platinum are to be removed from *clean scrap*.
The gold to be refined is fused with three times its
weight of copper or clean scrap silver, on charcoal,
if small in quantity, otherwise in a clay crucible. The
alloy obtained is next hammered or rolled out, cut in
small pieces and subjected to the action of strong
nitric acid (commercial) until entirely disintegrated.
When this point is reached, water is added to weaken
the acid somewhat, so that it will not attack the
filter and the residue of metallic gold is separated
from this solution by filtration, care being exercised
to wash with hot water until no acid is detected in
the washings.*   Both paper and gold should next be
dried and treated as already indicated in Method II.
Gold thus prepared is usually 996–998 fine.

---

*In case silver has been used to alloy the gold, save the
filtrate and the washings, and recover the silver as indicated on
page 130.

Refining Silver.

The problem here presented involves the preparation of pure silver for use in making amalgam-alloys, etc., from the following classes of scrap.

Class I. "*Standard*" *scrap* and other commercial alloys rarely containing, besides silver, a wider range of metals than copper and zinc. This class of scrap can be treated directly as indicated below.

Class II. Waste dental amalgam containing silver, tin, mercury and at times gold, platinum, zinc, copper, etc. First, separate the mercury as follows: Place the amalgam in an iron retort (Fig. 22), and apply sufficient heat to distill the mercury, which subsequently should be purified as shown on page 131. Remove the residue from the retort and refine as indicated below.

The recovery of silver from any alloy containing an abundance of tin is somewhat troublesome owing to the fact that after digesting the alloy in nitric acid, tin is converted into an insoluble residue which clogs a filter paper and prevents the silver solution from running through. These facts must be borne in mind in refining this class of waste, and instead of attempting to remove any bulky residue by filtration it should be allowed to settle and the clear liquid decanted or siphoned off. Finally, the residue should be thrown upon two folds of loose muslin attached to a wood frame and washed free from silver, i. e., until the washings give but a faint opalescence with hydrochloric acid or salt.

Place the metal in a flask, and cover with nitric acid. Apply heat to the flask placed on the sand bath, and at times add small quantities of acid to maintain the reaction. This operation should be carried on in a good draft or in a hood to carry off disagreeable fumes. When action finally ceases, dilute the acid somewhat with water, separate any insoluble residue and add to the clear liquid sufficient common salt solution to completely precipitate the silver as silver chloride. Stir vigorously, warm and let settle. Decant the clear liquid, and pass in water from a faucet in such a manner as to thoroughly agitate the precipitate. Repeat this operation two or three times, and finally transfer the precipitate to a wet filter and wash until acid is no longer detected in the washings. Remove the precipitate and paper from the funnel and dry, as in the case of gold, but do not try to incinerate the paper. Place paper and precipitate in a graphite crucible and cover with sodium carbonate or charcoal and potassium carbonate. Reduce the silver chloride to metallic silver in the furnace and pour into a clean bullion mold (Fig. 18), or granulate by pouring into water. Silver carefully prepared as indicated is pure enough for all ordinary purposes.

Refining Mercury.

The mercury to be refined may contain :

I. Mechanical impurities, as dust, etc. Refine as indicated in Method I.

II. Metallic impurities, as those retained when mercury is squeezed or distilled from amalgams.* Remove the bulk of the impurities by redistillation or by Method I., and finally refine by Method II. or III.

METHOD I.

Make a pin hole in blotting paper or in a rough filter paper placed in a funnel and filter the mercury. Much of the impurities will be left on the paper. This treatment, followed by washing, will usually be found sufficient to remove mechanical impurities.

METHOD II.

Place the mercury in a shallow dish and cover its surface with *very* dilute nitric acid. Agitate frequently during several hours. The acid will dissolve the impurities together with a little mercury. Finally wash the surface of the mercury under a stream of water from the faucet, dry with a clean sponge or blotting paper then on a water bath or over a low flame.

METHOD III.

Place the mercury in a dry bottle and cover its surface with finely powdered sugar. The mercury should occupy but one-fourth the capacity of the bottle. Shake the bottle vigorously, removing the cork at times to admit a fresh supply of air. The foreign metals are oxidized and cling to the grains of

---

*A portion of the foreign metals usually passes over with the mercury as vapor or as the result of spirting.

sugar. Filter as indicated in Method I., wash and dry. Pure mercury does not tarnish, has no film on its surface after standing, and does not leave a " tail " when made to run down a slight incline.

CHAPTER XIII.

DENTAL AMALGAMS AND AMALGAM=ALLOYS.

Certain alloys, commonly known as *dental amalgams*, are employed for filling cavities in decayed teeth. From the broad definition of the term given in Chapter X., it would follow that any metallic compound containing mercury is an amalgam. In dentistry, however, the expression is used in a restricted sense, having reference only to the product resulting from the admixture by direct contact of a finely divided metal or alloy with sufficient mercury to form a plastic mass. As just stated, amalgams are sometimes made by amalgamating a single metal. Thus precipitated copper (see p. 31), when triturated in a mortar with dilute mercuric nitrate and mercury, yields the so-called copper amalgam, the use of which is now practically discontinued. In most cases, however, amalgams for fillings are made from alloys composed chiefly of silver and tin, and commonly called *amalgam-alloys.* By combining these metals in varying proportions it is possible to produce alloys which yield more satisfactory amalgams for dental purposes than does a single metal, such as silver or copper or any combination of other metals. Silver-tin amalgams, however, possess certain objectionable features, and for the purpose of counteract-ing these, small quantities of some other metal or metals are often introduced. The metals usually added for this purpose are gold, copper, platinum, zinc, and occasionally aluminum and cadmium. As a rule, the proportion of silver and tin entering into these alloys varies from sixty-five per cent of silver and thirty-five per cent of tin to forty per cent of silver and sixty per cent of tin. , In certain cases, however, as high as seventy-four per cent of silver is added. The composition of many of the well-known amalgam-alloys can be seen by referring to the Appendix, Section II.

After being melted these alloys are cast in an ingot mold and are then reduced to a fine state, either by filing them or by cutting them in a lathe. In this con-dition they are intimately mixed with mercury by kneading in the palm of the hand or in a small mortar until a homogeneous, coherent mass has been formed. An amalgam which probably possesses definite chem-ical composition in many cases now exists dissolved in an excess of mercury, and to use it as a filling material it is necessary to remove the excess. Usu-ally this is accomplished by wringing through chamois skin with the aid of pliers. The mercury passes through the pores of the straining material, leaving a plastic, crystalline mass, which emits a peculiar crack-ling when pressed, probably due to the grating of the crystals upon each other. In this state the amalgam is packed in the cavity of the tooth, where, in a short time, it loses its plasticity and forms a hard, metallic plug.

Properties of Amalgams.

Amalgams possess many peculiar properties, and under certain conditions undergo changes which require special consideration.

COLOR AND LUSTER.

Amalgams are uniformly white or gray, even when gold or copper is present in large quantities. After hardening they are capable of taking a high silver-like polish and of retaining it, as well as their color, indefinitely in the absence of tarnishing substances. When placed in the cavity of the tooth, however, they sooner or later lose their brilliancy and discolor superficially. These changes, which are due chiefly to the influence of hydrogen sulphide resulting from the decomposition of certain foods in the mouth and somewhat to the action of drugs and vegetable acids, constitute an objection to amalgams as a filling material. Moreover, the sulphides and other compounds formed often permeate the tooth substance and greatly discolor it, a fact which is particularly noticeable in the case of silver-tin amalgams containing copper and those containing cadmium. According to many authorities, however, the substances formed, particularly in the case of silver and copper, exert a preservative influence upon the tooth.

Although the discoloration of amalgams cannot be prevented entirely, it can be controlled somewhat by eliminating copper from the formula of an amalgam-alloy and by adding a small proportion of gold or of zinc.

SOLIDIFICATION.

After an amalgam has been mixed it begins to lose its plasticity and in a comparatively short time it reaches a state of hardness in which it can no longer be worked. In dentistry this change is called "setting." It must not be inferred, however, that an amalgam has reached its hardest state just after solidi- fication has occurred. Indeed, it is generally recog- nized that the process of hardening continues for hours or even for days.

Amalgams differ greatly in the time required for solidification. Tin amalgams and those prepared from amalgam-alloys containing a high percentage of tin set slowly and imperfectly, while those amalgams formed from amalgam-alloys containing a large pro- portion of silver set very quickly and become very hard. The quickest setting are those prepared from alloys containing seventy to seventy-five per cent of silver.

Metals which tend to hasten setting when added in certain proportions to a silver-tin alloy are gold, platinum, copper, zinc and cadmium. Concerning the effects of the two first named metals, there seems to be considerable controversy. It is generally recog- nized, however, that the addition of platinum to a silver-tin alloy containing gold induces the property of setting quickly.

The proportion of mercury in an amalgam is a factor as regards solidification. An amalgam contain- ing an excess of mercury will set more slowly than the same amalgam with the excess of mercury removed.

Finally, the treatment to which an amalgam-alloy has been subjected after cutting and before amalgamating greatly affects the setting properties of the amalgam. A freshly cut alloy will yield a much more rapidly setting amalgam than will an "aged" or "annealed" sample of the same alloy, i. e., an alloy which in the form of filings or shavings has been exposed for a long time to the action of the air or has been heated for a short time at the temperature of boiling water.

CHANGE OF VOLUME.

At the time of the change from the plastic to the hard state and in some instances after hardening apparently has taken place, amalgams undergo a change of volume which in magnitude, in the time required for its completion, and in the manner in which it proceeds differs widely in different alloys and in different conditions of the same alloy. With most amalgams the change is so slight that it requires a delicate micrometer to detect it; in a few instances, however, it can be easily observed by the naked eye. In some cases it continues for a short time and then ceases; in others, it proceeds for hours or days and sometimes for weeks. Finally, it may result either in expansion or contraction. It does not always happen, however, that the final volume is the result of an increase or decrease alone. Indeed, it is quite common for an amalgam to expand for some time and then to contract or *vice versa*. Silver-tin alloys containing less than fifty per cent of silver produce amalgams which shrink at first and then expand. Those containing over seventy-five per cent of silver yield amalgams which expand only.

This tendency to change in volume constitutes the greatest objection to amalgams as filling material. An amalgam which contracts will draw away from the wall of the cavity and allow the ingress of moisture, i. e., "leakage." Those which expand will often assume a convex form on the surface when packed in a cavity and produce the phenomenon commonly termed "spheroiding." This tendency to " spheroid " is undoubtedly the result of expansion in the direction of least resistance and is similar to the phenomenon observed when water freezes in a vessel.

According to the researches of G. V. Black, expansion and contraction are influenced by the composition of the amalgam-alloy.

Of the silver-tin alloys, those containing sixty-five per cent of silver and thirty-five per cent of tin are said to yield amalgams which show the least expansion or contraction when used freshly cut.

An alloy containing seventy-three per cent of silver and twenty-seven per cent of tin produces, when fully annealed, an amalgam which shows no change of volume. Those metals which, when added in small proportions, are said to decrease the tendency to contract are gold and copper. According to the experiments of Dr. Black, zinc induces an expansion which continues for an indefinite time. A great diversity of opinion still exists as regards the effects upon amalgams of many of the metals commonly added in small quantities to amalgam-alloys.

The chemistry underlying this change of volume cannot be explained. It is interesting to note, however, that in one case, at least, it is the direct result of the oxidation of one of the constituents of the amalgam-alloy. A silver-tin alloy containing as little as one or two per cent of aluminum expands greatly and becomes hot when amalgamated in the hand. When the mass is placed under the microscope bubbles of gas are seen to escape. The explanation of these phenomena (deduced from experiments made by the author) is as follows: The aluminum amalgam in the presence of moisture decomposes the water, the aluminum being oxidized and the hydrogen of the water being set free. The hydrogen—some of which is seen to escape—being liberated within the mass naturally brings about a change analogous to that produced by carbon dioxide in raising bread. This explanation is substantiated by the fact that if the amalgam is made in a perfectly dry mortar comparatively little expansion takes place. If, however, a drop of water is added, the mass almost instantly increases to three or four times its original volume.

STRENGTH AND CHANGE OF FORM WITH PRESSURE.

A solid amalgam presents a peculiar combination of properties which seldom is observed in any one substance. Thus, if an amalgam is struck a sudden blow with a hammer, it will fly in pieces; but if the force applied is comparatively light it will yield greatly before breaking. In many instances it is possible to reduce an amalgam to a thin sheet. This tendency to yield to pressure is, in dentistry, commonly called "flow;" and the amount of pressure which an amalgam will sustain without breaking is taken as representing its strength. The flow of amalgams differs, however, in a certain respect from that of the metals. For example, a block of silver, steel, iron or gold when subjected to pressure yields suddenly and then ceases unless the pressure is increased. With a block of amalgam under like conditions a different phenomenon is observed: "When the flow has begun it continues as long as the stress is maintained. No increase of the stress is required to maintain the flow even after the area of the amalgam has been greatly increased by the flattening of the mass between plane surfaces. If a stress of fifty pounds be put upon a block of amalgam and maintained for one hour, flow will occur at a certain rate; if the stress be reduced to twenty-five pounds the flow will continue, but at a reduced rate." "It will go slowly with a light stress, somewhat quicker with a heavier stress, but it cannot be made to go very quickly with a very heavy stress; it will break into fragments."*

The flow of an amalgam is generally regarded as of greater importance than the crushing-strength, since many amalgams which are capable of withstanding the crushing force of mastication will gradually yield, and thus their adaptation to the margins will be destroyed. An increase of tin in a silver-tin alloy increases the flow and decreases the

---

*G. V. Black, *Dental Cosmos*, Vol. XXXVII., p. 558.

strength. The addition of but small proportions of gold to a silver-tin alloy greatly increases the flow and reduces the strength. With copper, the opposite is noted. Obviously the manner of mixing and packing, the time given for hardening, etc., have an influence upon the flow and strength of an amalgam, and for this reason the results of measurements of these properties on the same amalgam are not liable to be very constant.

SOLUBILITY.

Most amalgams are but slightly soluble in the fluids of the mouth, except when the saliva is strongly acid or alkaline and when other metals, such as gold and aluminum, are present. Under these conditions galvanic action is induced, and this greatly facilitates their solution. From the strong galvanic action often detected in the mouth, it would seem probable that the lack of permanence of fillings which is commonly attributed to the flow and to the expansion and contraction of amalgams is due in part to this agency. The chief objection to the use of copper amalgams is that they disintegrate under the action of the fluids of the mouth.

Preparing and Testing Amalgam-Alloys and Amalgams.*

It will be the object here not only to teach the mechanical manipulation involved in making these

---

*A convenient form of record to be kept by the student in performing these exercises is shown in the Appendix, Section II.

alloys, but also to offer some opportunity to observe their physical properties. Since this can best be accomplished by laboratory exercises, the student is directed to prepare about twenty-five grams of amalgam-alloy according to a formula to be assigned by the instructor, and to amalgamate it and to study and test the resulting amalgam as described later. In doing this work the metals may be used which have been refined as directed in Chapter XII.

The general directions to be followed and precautions to be observed in making alloys have already been given in Chapter X., and are applicable here to a certain extent. The preparation of amalgam-alloys, however, involves certain operations not described elsewhere in this book, and in order that the student may have a detailed outline to follow in preparing these special alloys, the complete process is described. The various steps consist in:

1. Weighing the metals.
2. Melting and pouring.
3. Comminuting. .
4. Sifting and removing particles of iron.
5. Annealing.

1. Weighing the Metals. First, calculate from the formula of the alloy the quantity of each metal needed to make the required quantity of alloy, and then proceed to weigh out the metals, using for this purpose the balance illustrated in Fig. 4. For convenience in handling the metal should be in the form of small pieces. Silver can be purchased in the granulated and tin in the shot form.

2. *Melting and Pouring.* Select a clean graphite or clay crucible and add the silver, covering its surface with borax. Place the crucible in the furnace and heat until the silver melts and sinks below the surface of the molten borax. Without removing the crucible from the furnace introduce the tin (also gold and copper). When the content of the crucible is in a molten state, stir carefully once or twice with a pine stick (do not use an iron rod, as is often directed). When satisfied that the metals are thoroughly alloyed remove the crucible from the furnace, give it a gentle circular motion with the tongs and pour the alloy as quickly as possible into a clean mold (Fig. 18).. When zinc or cadmium is to be added, introduce it with constant stirring after removing the crucible from the furnace.

3. *Comminuting.* Amalgam-alloys are converted into a fine state either by filing them or by cutting them in a lathe. For filing an alloy the so-called bastard file, about fourteen inches in length, is best suited. When alloys are to be converted into shavings in a lathe they should be cast in a mold which furnishes a rod-shaped ingot. An alloy containing a large percentage of tin will be found to cut easily and to give coarse filings. Such alloys have a tendency to clog the file and hinder its action. In case this happens clean it with a steel brush. Alloys rich in silver give fine filings, and hence the character of the filings is often taken as indicating the approximate composition of the alloy.

4. *Sifting and Removing Particles of Iron.* When

an alloy has been filed, it should be sifted through a sieve of forty, sixty or eighty meshes, depending upon the character of the filings, to remove pieces of alloy which have been broken from the ingot.

After sifting, an ordinary magnet should be passed through the filings in order to remove particles of iron which have been broken from the file.

5. Annealing. Attention has already been called to the fact that filings which have been cut for some time or have been heated possess certain properties not shown by those freshly cut. This peculiar change, which undoubtedly is due in part at least to a superficial oxidation of the tin since in the alloyed state this metal is very susceptible to oxidation, is looked upon as conferring desirable properties upon the mass when amalgamated. There is considerable controversy concerning the chemistry of annealing, some insisting that it is a molecular change. Whatever importance may be attached to this theory, it nevertheless is true that in many respects the theory of superficial oxidation is consistent with the facts observed.

To anneal an alloy, place the filings in a test tube or flask and keep the vessel in boiling water for fifteen minutes. This time is usually sufficient, although in some cases in order to obtain the desired results it may be necessary to heat for a longer time. It is a noticeable fact that when annealing is carried too far the amalgam sets slowly and is reduced in strength. Thus, if an alloy is heated a minute or so in the naked flame at a temperature somewhat above boiling water, it will turn brown, due to the oxidation of the tin, and will produce an amalgam which is practically worthless. Annealed alloys take up much less mercury than freshly cut and yield a greater quantity of the so-called "dirt," upon mixing. This black substance is a lower oxide of tin.

MIXING AMALGAMS.

The amalgamation of filings or of shavings is usually effected by rubbing them with mercury in the palm of the hand, in a small mortar or in both. In all cases the mixing should be thorough and the mercury used should be pure. The quantity of mercury to be employed necessarily varies with different alloys and with the different conditions of the same alloy. Freshly cut alloys require much more mercury than those which have been annealed, and fine filings take up more mercury than those which are coarse. The proportions usually range, however, from two parts of alloy and one of mercury to equal parts of each.

It is often urged that in mixing amalgams the proportion in which the mercury and the alloy combine be determined by experiment and that thereafter they be weighed out accurately in that proportion. This method is seldom followed in practice. In all cases, however, if the constituents are not weighed care should be taken to avoid a great excess of mercury. The chief reason for this arises from the fact that in expressing the excess of mercury from an amalgam more or less tin is removed, and thus the composition of the amalgam-alloy is changed. To what extent comparatively large proportions of mercury may affect the composition of an amalgam-alloy can be seen by briefly reviewing some results obtained by the author. A silver-tin alloy of the composition, silver sixty-five, tin thirty-five, was made and a certain weight of this alloy in the form of filings was mixed with an equal weight of mercury. The excess was removed by squeezing through heavy muslin, and the expressed mercury was analyzed. In no case was more than a trace of the silver in the amalgam-alloy removed. With the annealed alloy 1.7 per cent of the tin was removed, and with the freshly cut 0.86 per cent was removed. A second series of experiments in which the same weight of alloy as above and twice the weight of mercury was used, gave surprisingly different results. Under these conditions 4.38 per cent of tin was removed from the annealed and 4.18 per cent from the freshly cut alloy. A third series of experiments in which the alloy and mercury were mixed in the proportion of one part of the former to five parts of the latter showed that 9.6 per cent of the tin was removed from the annealed and 9.8 per cent from the freshly cut alloy. An interesting fact developed by these experiments was that in all the determinations made, about twenty-five in number, the tin constituted almost exactly one per cent of the mercury removed.

In removing the excess of mercury from an amalgam, chamois skin or good muslin may be used. Obviously the pressure exerted in wringing will determine the proportion of mercury left in the amalgam, which in

all cases should be the least that will suffice to make
a mass which can be manipulated. Certain devices
are suggested for amalgamating by shaking. The
objection to using these, however, will become apparent if some alloy and mercury are shaken in a test
tube. The result will be that the black substance
commonly observed in mixing amalgams, i. e., the
lower oxide of tin, will separate in large quantities.
Thus it is seen that the danger of changing the composition of the amalgam-alloy by this method of mixing is very great. Indeed, it is possible by continued
shaking as described to remove a large proportion of
the tin from an amalgam.

TESTING AMALGAM-ALLOYS AND AMALGAMS.

With the ingot of amalgam-alloy which has been
prepared the following exercise may be performed
before it is converted into filings:

1. Determination of Specific Gravity. By means of
a hack-saw cut the ingot into three pieces of about
equal size and mark each by a scratch with a file.
Next weigh each portion of the ingot on the analytical balance and record the results as *weight in air.*
Then place a small bench, which is to be found in
the drawer of the balance case, astride the left-hand
scale pan in such a manner as to allow the pan to
swing freely below it. In this place a 200 c. c. beaker
two-thirds filled with distilled water at about 15° C.
Next tie a piece of linen thread about eight inches
long to one of these ingots and suspend it from a
hook on the stirrup, carrying the left-hand scale pan,

in such a manner that the ingot is completely immersed in the water in the beaker. By means of a fine brush remove any bubbles of air that may cling to the ingot, and then counterpoise it by placing the required weights on the right-hand scale pan. Record the result as the *weight in water*. With the weight in air and in water known, the specific gravity of the portion of the ingot in question can be determined by the following formula:

$$\frac{\text{Weight in air}}{\text{Weight in air—Weight in water}} = \text{Specific gravity.}$$

In this manner the specific gravity of all three portions should be determined and compared. In case the ingot is homogeneous these values should agree very closely. If, however, from any reason the metals have not been properly combined, more or less variation will be observed. After these facts have been noted, the average of the three results may be compared with the specific gravity calculated from the proportions of the metals employed in preparing the alloy. In this way the change in specific gravity accompanying alloying may be noted.

The three portions of the ingot may now be converted into filings which may be employed for the following tests:

2. Test for Discoloration. Amalgamate about two grams of the alloy, roll it into a ball and place it in a test tube. Fill the tube with distilled water, saturate with hydrogen sulphide gas, and cork the tube. After twenty-four or forty-eight hours the color of this sample may be compared with that of some freshly mixed amalgam. In this way the tendency of different amalgams to discolor can be roughly determined.

3. Tests for Change of Volume. Before making these tests anneal about one-half of the amalgam-alloy which has been prepared and place the annealed sample in a labeled bottle.

I. Weigh out accurately two grams of the freshly cut alloy on the analytical balance. Remove the weights from the scale pan, place a gram weight on the pan with the alloy and exactly counterpoise the alloy and gram weight with mercury. This gives two grams of alloy and three grams of mercury. Transfer both the alloy and the mercury from the scale pan to a rubber cot. Amalgamate thoroughly, squeeze out the excess of mercury, carefully collect this on a sheet of filter paper and retain it in a labeled test tube for use in *Special Tests* given later. Pack the amalgam in one of the cavities in the steel block* (Fig. 29) with filling instruments. Dress the surface of the filling level with the face of the block (do not burnish) and take particular care so to adapt it to the margins of the cavity that when viewed under the microscope no openings are observed.

II. In the manner just described prepare a sam-

---

*Before making these tests it will be necessary for the student to burnish the faces of the steel block, and carefully to smooth the margins of the cavities so that they show no irregularities under the microscope.

ple of amalgam from the annealed alloy, weighing the constituents, amalgamating and retaining the expressed mercury in a second labeled test tube. Pack the resulting mass of amalgam in the second cavity of the steel block, following the directions given above.

III.   Finally prepare a third sample of amalgam as follows: Weigh out as directed two grams of the freshly cut alloy and three grams of mercury.   Place the alloy in a dry test tube or porcelain crucible and subject it to a low heat over a Bunsen flame until somewhat oxidized, i. e., until it turns brown. Amalgamate this, retain the expressed mercury in a third labeled test tube and pack the mass in the third cavity of the steel block.

When the fillings have been completed, the steel block may be placed in the *micrometer* and the change of volume measured from time to time as directed on page 116. Microscopical examinations of the margins of the fillings should accompany the micrometrical measurements. In all this work the student should observe the general working properties of the different samples of amalgam, particular attention being directed to their setting properties.

4. Tests for Strength and Flow. Prepare two cubes of amalgam from the freshly cut alloy by packing them in the matrix.   (Fig. 31.) When they have been given five days in which to harden determine their strength and flow in the *dynamometer* as described on page 118. Prepare and test in the same manner cubes of amalgam made from the annealed and from some oxidized alloy and note the effects of annealing and oxidizing upon strength and flow.

5. Special Tests. In the study of amalgams it is essential that some tests be made to show the different quantities of mercury taken up by equal weights of the same amalgam alloy in the freshly cut, in the annealed and in the oxidized states, and to determine the proportions of tin* removed in squeezing out the excess of mercury. In making these determinations the mercury which was retained in the labeled test tubes from the *Tests for Change of Volume* may be used.

I. To determine the different quantities of mercury taken up under the conditions stated above: Weigh the contents of each test tube. Since three grams of mercury were employed in each case, the mercury retained by each amalgam is equal to the difference between this weight and that of the expressed mercury.† *Retain the three samples of expressed mercury for tests given below.*

Problem 1. Knowing the formula of the amalgam-alloy, the weight of amalgam-alloy taken (2 grams) and the quantity of mercury in each amalgam, calculate the percentage composition of the three amalgams which were tested for change of volume.

---

*Although a very small quantity of silver may be removed in the mercury expressed from a silver-tin alloy, it would be impracticable for the student to attempt to estimate it. Whether such metals as zinc and aluminum are removed, if present in the amalgam-alloy, has not been definitely determined.

†This result, of course, will not be exactly correct, since the weight, owing to the presence of more or less tin, does not represent pure mercury.

II.   To determine the quantities of tin removed from the amalgam-alloy by the three samples of expressed mercury proceed with each as follows: Carefully transfer the mercury from the test tube to a 200 c. c. beaker (the beaker should be labeled); add about 10 c. c. of concentrated nitric acid and an equal volume of distilled water; warm gently over a hot-plate. The mercury dissolves and the tin, if present, remains as a white residue, SnO_2. When the mercury is dissolved, boil the contents of the beaker gently for five minutes, then add 25 c. c. of distilled water and filter through a quantitative filter paper.* Wash the residue on the paper until no acid can be detected in the washings.   Reject both the filtrate and washings and allow the paper and residue to dry in an air bath. When dry, wrap the paper containing the residue into a small bundle and place it in a porcelain crucible (Fig. 26) the weight of which should have been determined previously.   Place the crucible on a pipestem triangle (Fig. 27) and apply the flame of the Bunsen burner until the paper is completely burned and the residue is white. Allow the crucible to cool and then weigh. Subtract the weight of the crucible from the combined weight of the crucible and the residue and the result is the weight of the residue, which is tin oxide, SnO_2. Multiply this by the factor 0.788 and the result is the weight of metallic tin removed by x grams of mercury.

---

*A special kind of paper used in quantitative analysis which gives practically no ash when incinerated.

Problem 2. Determine what percentage of the x grams of expressed mercury is tin.

Problem 3. Determine what percentage the tin removed is of the total weight of tin in the two grams of amalgam-alloy taken.

After completing the various tests outlined above the student should test in the·same manner the amalgams made from one or more of the prominent amalgam-alloys found upon the market.

Copper Amalgam.*

Owing to the fact that copper amalgam has practically gone out of use in dentistry, it is but briefly considered here. To prepare a small quantity as a laboratory exercise the student may proceed as follows : Weigh out roughly twenty-five grams of pure copper sulphate crystals and dissolve, with the aid of heat, in 250 c. c. of distilled water. When the crystals are dissolved acidify the solution with sulphuric acid and immerse in it a rod of iron or of zinc. The precipitation of red metallic copper begins at once and continues until it is removed from the solution, i. e., until the blue color of the solution has disappeared. The bar of iron or of zinc is freed from adhering copper, then removed; the clear supernatant liquid is poured off; the residue of copper is washed by decantation several times with hot water containing

---

*For more facts concerning copper amalgam than are given in this book, the student is referred to Flagg's ''Plastics and Plastic Filling'' and to the article by G. V. Black in the *Dental Cosmos*, Vol. XXXVII., p. 737.

a little sulphuric acid, and finally with cold water until no acid is detected in the last washings. The copper is next transferred to a mortar and moistened with a solution of mercuric nitrate, which forms a coating of mercury on it. About twelve or thirteen grams of mercury are then added and rubbed with the copper until amalgamation is completed. The mass of amalgam is next washed with water and squeezed in chamois skin or muslin. Again it is rubbed in the mortar, squeezed in chamois skin and finally made into pellets and allowed to harden.

When required for use a pellet is placed upon a spatula and heated over a flame until the "beads" of mercury appear on the surface. *The heating must not be carried too far, for this will oxidize the copper.* It is then crushed in a mortar until plastic, in which condition it is ready for use.

According to the tests of G. V. Black, copper amalgam retains its margins well, changes in volume but slightly and furnishes a very rigid filling material. These tests further show that copper amalgam is reduced in strength by frequent reheating, and hence the residue amalgam should not be used.

CHAPTER XIV.

THE ASSAY OF AMALGAM-ALLOYS.*

By the application of certain chemical methods it is possible to. determine accurately the proportions of the constituents present in a complex substance such as an alloy. This is known as *quantitative* chemical analysis as distinguished from *qualitative* which, as has already been shown,† serves only to separate and identify the constituents without determining their quantity. Quantitative analysis may be divided into *gravimetric* and *volumetric.* In gravimetric analysis the substance is usually precipitated in some insoluble form much as in qualitative and then carefully filtered, dried and weighed. The process followed in volumetric analysis consists in estimating the substance by measuring accurately the volume of a reagent of known content, commonly called a *standard solution,* necessary to bring about a certain complete reaction. The determination of tin and zinc as outlined hereafter is an example of gravimetric analysis, while the estimation of copper by means of a solution of potassium cyanide illustrates volumetric.

The quantitative analysis of many amalgam-

*The form of the report to be submitted upon the completion of this work is shown in the Appendix, Section II.

†See Chapter IX.

alloys is not difficult and some knowledge of simple
methods may at times be a matter of importance to the
dentist. To determine successfully the composition of
alloys containing a greater variety of metals than
silver, tin, copper, zinc and gold requires an extensive knowledge of analytical chemistry and hence
is not touched upon here.

The student may apply the methods outlined hereafter to the assay of a sample of the alloy used in the
work upon amalgams, and in so doing he can determine how nearly it is possible to realize in the ingot
the proportions of the metals placed in the crucible.
The general scheme of separation is given in the
accompanying table and the details· of manipulation
are explained under the different methods.

Method I.

ESTIMATION OF TIN.

The residue remaining on the filter after dissolving the sample in nitric acid consists of the tin as
oxide, SnO_2, and the gold.   If the latter is present
it will cause the precipitate to be purple in color.
When gold is present it is weighed with the tin oxide
and from the weight of the two the amount of gold
found by the *Fire Assay* (Method IV.) is deducted and
the tin oxide equals the difference.

Place the dried precipitate in a weighed porcelain
crucible and ignite gently at first, then at the highest
heat of the Bunsen burner.   Cool and weigh.   The
weight of SnO_2 multiplied by 0.788 equals the
amount of tin present.   This multiplied by 100 gives
the percentage of tin in the alloy.

THE ASSAY OF AMALGAM-ALLOYS

Containing Silver, Tin, Copper, Zinc and Gold.

One gram is accurately weighed out on a watch crystal and brushed into a 200 c. c. beaker. Add a mixture of 10 c. c. strong nitric acid and 10 c. c. water and place on the warm part of a hot-plate until action ceases and then evaporate to "moist" dryness. Add distilled water and a few drops of nitric acid; stir, warm and allow any residue to settle. When settled perfectly clear, filter on a quantitative filter paper, and wash the filter with hot water until the washings give no acid reaction with litmus paper.

Residue.

Contains **tin** as oxide, SnO_2, and the **gold.**

For the estimation of tin see Method I.

The gold is determined by fire assay. See Method IV.

Filtrate may contain Silver, Copper and Zinc.

Transfer the filtrate to a porcelain dish and evaporate to "moist" dryness on a water bath or on the warm part of a hot-plate. Then wash down the sides of the dish with hot water and add a few drops of strong hydrochloric acid to precipitate the silver as silver chloride. Make sure that the silver is all precipitated by adding more hydrochloric acid and stirring to collect the precipitate. Filter and wash thoroughly with hot water containing a few drops of nitric acid. Reserve filtrate and washings in the same beaker.

Precipitate.

Silver chloride, AgCl, Reject.

For the determination of silver see Method IV.

Filtrate may contain **zinc** and **copper.** The presence of the latter may be detected by the blue color.*

Warm the filtrate and add hydrogen sulphide gas until saturated. Filter, washing precipitate with water containing hydrogen sulphide gas.

Precipitate.

Copper sulphide, CuS, black. See Method II.

Filtrate may contain **zinc.** Boil with a little potassium chlorate, and filter out free sulphur and proceed as in Method III.

*In case no copper is detected by the blue color, the filtrate may be treated at once for zinc.

Method II.   .

ESTIMATION OF COPPER.

The estimation of copper may be accurately and quickly made by the so-called *volumetric cyanide method.* It requires a solution of potassium cyanide, the equivalent of which in copper is carefully determined.*

Dissolve the precipitate of copper sulphide on the paper with hot nitric acid and wash the paper with hot water, collecting the washings in the beaker with the nitric acid solution. If the sulphur residue on the paper is dark colored then wash all the contents of the filter into the nitric acid solution and boil. Filter and wash. Cool the filtrate, add a slight excess of ammonium hydroxide and then add the potassium cyanide from a burette (Fig. 35) to the copper solution in a beaker until the blue color disappears. Multiply the number of cubic centimeters of cyanide solution used by the weight of copper to which 1 c. c. of the cyanide solution is equivalent, then multiply the product by 100, which gives the percentage of copper in the alloy.

Method III.   .

ESTIMATION OF ZINC.

Dilute the filtrate, which may contain zinc, to about 400 c. c., and heat it to boiling. Now remove the beaker from the hot-plate, cover with a watch crystal,

---

*For directions for preparing and standardizing this solution see Appendix, Section I.

and add from the end of a spatula, a little at a time, sodium carbonate, replacing the watch crystal after each addition to prevent loss by spirting. Continue the addition of the sodium carbonate until the zinc is all precipitated and the sodium carbonate is in excess, which may be ascertained by a piece of litmus paper. Now place the beaker on the hot-plate, heat it to boiling, stirring at intervals to prevent bumping. Boil hard for about fifteen minutes. Remove the beaker from the hot-plate and let it stand until the precipitate is completely settled. Decant the supernatant liquid onto a quantitative filter, and to the precipitate add about 100 c. c. of hot water and stir thoroughly. Allow to settle and again decant the clear supernatant liquid. Repeat the washing by decantation about five times, then wash the precipitate onto a filter and continue washing the precipitate with hot water until the wash water shows no alkaline reaction with red litmus paper. Allow the filter to drain, then transfer both precipitate and paper to a weighed porcelain crucible. Ignite very cautiously at first to avoid spirting. This may best be done by playing the flame on the crucible until the precipitate is dry. Finally heat at a high temperature until the paper is burned and the residue is light yellow in color. Cool and weigh. Weight of zinc oxide, ZnO, multiplied by 0.803 gives the weight of zinc, and this result multiplied by 100 gives the percentage of zinc in the alloy.

Method IV.

FIRE ASSAY FOR GOLD AND SILVER.

The assay of alloys for gold and silver by ordinary gravimetric or volumetric methods is not as quick nor as accurate as the fire methods, and this is particularly true in alloys containing much tin or copper.

The fire assay consists in melting the alloy in twenty to forty times its weight of granulated lead contained in a scorifier (Fig. 13) together with a small quantity of borax glass. The latter being a strong flux unites with the base metals to form a slag or glass which is essentially borates of lead and of other base metals. But as the amount of borax used is small there necessarily will be but a small slag which at first appears as a glassy ring in the outer edge of the molten metals. If a current of air is allowed to pass over the hot lead (and other metals) it is oxidized and passes off as fumes of lead oxide. If now the heat be continued the amount of lead will keep decreasing and the ring of slag will appear to become larger; but in reality it is only "closing in" as the mass of lead becomes smaller from volatilization. On heating long enough, which usually is from thirty to forty minutes, the slag will close in and cover the whole surface of the lead, leaving but a small button hidden from view in the bottom of the scorifier. As gold and silver do not form compounds with the borax, but have a strong affinity for the hot lead, the button will contain all the gold and silver with possibly small amounts of copper or tin.

On cooling and separating the slag from the button the next operation is to separate the gold and silver from the lead. This is done by again melting the lead, this time in a cupel (Fig. 15), allowing a current of air to pass over it. The red hot lead is converted into lead oxide, which is partly absorbed by the cupel and partly volatilized. The operation is continued until finally all the lead, with traces of copper and tin, are driven off or absorbed and nothing remains but the bright bead of gold and silver.

The bead is then weighed, the silver dissolved by nitric acid and the insoluble gold again weighed. This subtracted from the total weight of gold and silver gives the weight of silver.

From the foregoing it will be seen that the assay really consists of several operations which are followed in regular order :

1. Weighing sample and preparing charge.
2. Scorification.
3. Cupellation.
4. Weighing gold and silver bead.
5. Parting.
6. Weighing and calculating.

1. Weighing Sample and Preparing Charge. Two assay tons (60 grams) of granulated lead are weighed out on the pulp balance and about one-half is placed in a scorifier. Before doing this, however, chalk the inside of the scorifier. This prevents the hot lead from attacking it and eating a hole through it. Now weigh out *accurately* on the analytical balance one gram of the sample alloy and carefully brush on top of the granulated lead in the scorifier, then add the balance of the lead. On top place a small pinch of borax glass equal to about one gram.

2. Scorification. Place scorifier and contents in the back part of the muffle, close the latter with the plug and increase the heat. When the lead is melted and the ring of slag is formed, draw the scorifier to the center of the muffle and allow the air to enter. The plug is not replaced unless the muffle should get too cool. It will now be noticed that the fumes of lead oxide are continually forming on the surface of the lead and passing off up the chimney. The heat is continued until the ring of slag gradually closes in and in from thirty to forty minutes completely covers the hot lead. Now replace the plug and heat strongly for about five minutes. Then remove the scorifier and pour the contents into a button mold (Fig. 17). It is a good plan to warm the mold before pouring. This may be done by placing it on the top of the muffle furnace during the operation of scorification. When cold, the slag and button are emptied from the mold and the button is hammered to free it from slag. It is usually hammered into a cube. The button should be malleable and not crack. If it does show a tendency to crack it may be due to copper or some other base metal. In such cases it is best to rescorify with about a ton and a half of lead and a pinch of borax glass.

3. Cupellation. Select a cupel of about twice the weight of the button and place it in the hot part of the muffle for a few moments. Next *carefully* place the button in the cupel and close the muffle with the plug. If the cupel and muffle are quite hot the lead will melt at once and the surface will be seen to brighten and fumes of lead will begin to come off. The cupel is now carefully drawn forward and the heat so regulated that the "scales" or "feathers" of crystallized lead oxide appear on the inside of the cupel. It is difficult to describe the proper heat that is required, but it may be said that it is about right when the "feather" forms. It is essential that the heat in cupelling be carefully watched. The reason for this is that silver is appreciably volatile at high temperature. On the other hand, the temperature must not be so low as to allow the cupel to "freeze" and oxidation to stop. Should the latter occur a piece of charcoal placed against the cupel will usually heat it sufficiently to again start oxidation. Whenever freezing occurs, however, the results are usually doubtful.

When the lead has disappeared by absorption and by volatilization and the bead of gold and silver appears, replace the plug, remove the cupel to the back of the muffle and heat for about five minutes to drive off the last trace of lead. Then turn off the heat, remove the plug and gradually draw the cupel to the front of the muffle to cool. Do not cool too suddenly or the bead will "sprout," and possibly occasion a loss. The next step is to weigh the gold and silver bead.

4. Weighing the Gold and Silver Bead. Carry the cool cupel containing the bead to the balance and with a pair of pinchers pick up the bead. Tighten the pinchers, and with a stiff brush clean the bead, then place it in the pan of the balance and weigh. Record the weight as total gold and silver. The next operation is to separate or "part" the gold and silver.

5. Parting. Pure silver is readily dissolved in nitric acid, but if gold be alloyed with it in considerable proportion its solubility is decreased so that to completely separate gold and silver by this means it is essential that they be in the proportion of about two parts of silver to one of gold. But as silver is always present in amalgam-alloys in very much larger proportions than gold it will not be necessary to "inquart." After weighing, place the bead on a clean anvil, and with a small hammer flatten it. If the bead is hard and not malleable it may be due to the presence of some other metal. This must be removed by wrapping in about five grams of pure sheet lead and again cupelling and weighing. Often the bead may be rendered more malleable by heating to a dull red heat on charcoal with a blowpipe for a few moments. But if the brittleness be due to the presence of a foreign metal it will be necessary to recupel or the result will be too high in silver.

Place the flattened bead in a porcelain crucible and fill half full of nitric acid, sp. gr. 1.2. Boil on a hot-plate, keeping the crucible covered with a watch crystal. After it has boiled for several minutes and the brown fumes have passed off remove from the hot-plate and cool. Then with the aid of a glass rod pour off the bulk of the acid and fill the crucible half full of nitric acid, sp. gr. 1.3, and again boil for a few moments to dissolve the last trace of silver. The gold will now be found in the bottom of the crucible as fine black particles, contaminated with some silver nitrate in solution. The latter is removed by washing with distilled water. This is done by filling the crucible with water, tapping *gently* to cause the particles of gold to settle and then pouring off the water, using a glass rod. Again add water, repeating the operation four or five times. Finally drain the crucible as much as possible, and then soak up the last drop of water with a bit of filter paper, but be *very careful* that the paper does not touch any particles of gold. Wipe the crucible dry with the paper and complete the drying by placing it on the warm part of the hot-plate. Next transfer to a triangle and heat for a few moments at a red heat, which will bring out the yellow color. If no gold is present the first acid treatment will dissolve the silver to a clear solution. But no matter how small the black, insoluble gold may be, an attempt should be made to weigh it. If, however, it can be seen but cannot be weighed on a sensitive balance it is then usually called a " trace."

6. Weighing and Calculating. When the crucible is cool enough to handle, it is taken to the balance and the gold transferred to the balance pan by the aid of a needle point and weighed. This weight gives the gold. Multiply by 100 to convert into percentage. Subtract the weight of gold from the total weight of gold and silver and the difference equals the silver. Multiply this by 100 to obtain the percentage of silver in the alloy. In case gold is found in the alloy its weight should be subtracted from the weight of tin oxide obtained under Method I. This result multiplied by the factor 0.788, and then by 100 gives the percentage of tin.

CHAPTER XV.

SOLDERS AND SOLDERING.

Solders are alloys used to join metallic surfaces. As a rule they consist of the metal upon which they are to be used alloyed with some other metal or metals capable of considerably lowering the melting point without greatly modifying other physical properties. When dissimilar metals are to be united, a solder should be used which possesses an affinity for both and corresponds as nearly as possible to them in color, hardness, malleability, etc. In all cases solders should flow readily. In addition, it is obvious that solders for dental purposes should discolor but slightly and should be capable of resisting the action of the fluids of the mouth.

At times it is necessary to unite metals without the use of solders. This is accomplished by a process known as *autogenous soldering*, which consists in fusing together the contiguous parts. This method of soldering is extensively employed in plumbing and in the manufacture of various apparatus for chemical industries. Thus lead chambers used in the manufacture of sulphuric acid are soldered autogenously, since a common soft solder containing tin would soon corrode. By using the oxyhydrogen blowpipe it is possible to solder platinum vessels for chemical purposes in this manner where a solder of even gold, with which platinum usually is soldered, could not be used to advantage.

Many kinds of solders, known by the names of *tin, aluminum, copper, brass, argentan, silver, gold, jewelers', plumbers',* etc., are used in the arts. Although the names just given are commonly employed in distinguishing one solder from another, all may be broadly classified as *hard* or *soft.* Hard solders comprise those which fuse at or above red heat, and hence are used on metals possessing high melting points, while soft solders include those which fuse easily, and are used particularly by plumbers and tinsmiths, on metals fusing at low temperatures.

Preparation of Solders.

In preparing solders few directions are necessary beyond those already given for making other alloys. When zinc is to be added to a solder it should be introduced after the other constituents are melted and after the temperature has been reduced somewhat, otherwise it will be partially or even wholly oxidized. When both copper and zinc are to be added to a solder it is advisable to use instead of the separate metals, the proper quantity of pure brass, since in this state the zinc is less liable to be oxidized.

Soft solders are commonly used in the stick form; hard solders are often cast into ingots and converted into filings; gold and silver solders are usually rolled into sheets and used in the form of clippings.

As a laboratory exercise* in solder making, the student may prepare such quantities of aluminum, silver and gold solders as he may require for use in other departments of work.

SOFT SOLDERS.

Tin Solders. Tin solders are used chiefly in soldering tin plate, copper and Britannia metal. In dentistry they are sometimes employed in making appliances for regulating teeth. The best solder of this class consists of tin, two parts, and lead, one part; the common variety is composed of tin, one and one-half parts, and lead, one part. The various tin solders and the temperatures at which they melt are given in the following table :

No.	Parts.		Melting Point.	No.	Parts.		Melting Point.
	Tin.	Lead.			Tin.	Lead.	
1	1	25	292° C.	7	1½	1	168° C.
2	1	10	283° C.	8	2	1	171° C.
3	1.	5	266° C.	9	3	1	180° C.
4	1	3	250° C.	10	4	1	185° C.
5	1	2	227° C.	11	5	1	192° C.
6	1	1	188° C.	12	6	1	194° C.

In preparing tin solders, melt the tin first and then add the lead; stir vigorously and pour into a cold mold. The melting should be done in a crucible in-

---

*For the form of report to be made upon this work see Appendix, Section II.

stead of an iron ladle, as a little iron absorbed by the solder would greatly increase its melting point.

Chemical Solder. Pure tin in the form of foil is often used in soldering small articles. The pieces are fitted together with the foil between them and then held in the flame of a Bunsen burner until joined. In this manner an invisible joint can be made if the soldering is conducted carefully.

Bismuth Solders. The so-called bismuth solders, which properly may be classed with the fusible alloys described later, are composed of bismuth, lead and tin, and melt at temperatures ranging from 95° C. to 160° C. They are very fluid when melted and considerably harder than common solders. Although very satisfactory they are too expensive for general use owing to the content of bismuth. The formulæ of some bismuth solders are given below:

No.	PARTS.			Melting Point.
	Tin.	Lead.	Bismuth.	
1	3	5	3	94.4° C.
2	2	2	1	109.4° C.
3	2	1	2	113.3° C.
4	1	1	1	123.3° C.
5	3	3	1	154.4° C.
6	4	4	1	160.0° C.

Aluminum Solders. For dental and other purposes aluminum possesses many advantages over other metals, and hence the soldering of it is a matter of great importance; but, although many solders and methods of soldering have been proposed, few if any of them have proved to be entirely successful. The obstacles encountered in soldering aluminum are many. It melts at a comparatively low temperature, and hence can be soldered only with an alloy possessing a low melting point. But to obtain a solder which will melt easily and at the same time produce a strong joint is a difficult matter. It must be composed chiefly of the more fusible metals, such as tin, zinc, lead, etc. These, however, form weak solders which seem unable to "wet" the aluminum, and hence to attach themselves firmly to the surfaces to be united. Again, for dental purposes a solder must be employed which will retain its color and will not dissolve in the fluids of the mouth. Finally, a flux must be employed which will clean and protect from oxidation both the metal and the solder without attacking either. Common fluxes, as borax, cannot be used for this purpose since they exert an injurious influence upon aluminum and prevent union.

The solders given below are suggested by Schlosser as particularly adapted to soldering dental work, since they resist the action of corrosive substances.

PLATINUM-ALUMINUM SOLDER.			GOLD-ALUMINUM SOLDER.		
Gold	3	parts.	Gold	5	parts.
Platinum	0.1	"	Copper	1	"
Silver	2	"	Silver	1	"
Aluminum	10	"	Aluminum	2	"

When silver chloride is fused on aluminum it is reduced and the silver forms an alloy on the surface

of the aluminum.   These facts are taken advantage of
for soldering aluminum.·   The fused, finely powdered
silver chloride is used as a flux with ordinary solders.
It is placed at the junction and the soldering completed with the brazing blowpipe.

A solder recommended by Richards consists of :

Tin........................... .....29 parts.
Zinc........................ 11   "
Aluminum....... 1   "
Phosphor tin (tin containing phosphorus)1   "

This solder fuses easily and may be applied with
a copper or nickel soldering iron. The edges to
be soldered are scraped clean and then tinned
either by heating and rubbing them with the solder
or by applying the solder to them with a soldering
iron.   When this is done the soldering can be completed in any desired manner without flux.

A solder which has proved very satisfactory for
general use in the dental laboratory is that prepared
by the author's students.   It consists of:

Aluminum .... 45 parts.
Tin...................... 45   "
Mercury 10   "

This solder may be applied by aid of a brazing
blowpipe and a piece of steel wire.

Although there may be solders and methods of
soldering more satisfactory than those given, it should·
be remembered that it is a difficult matter to learn
much about them, as they are held in great secrecy by
those employing them.

HARD SOLDERS.

Brass Solders. Brass solders are sometimes used in place of soft or silver solders, for soldering brass, copper, etc. They consist of copper, zinc and sometimes tin. As the content of tin increases, the color of the brass becomes lighter and its ductility is diminished. Brass solders containing no tin are yellow, while those alloyed with this metal are known as *half white* and *white*, as shown below. In the following table are given the proportions of brass, tin and zinc used in making some common brass solders:

	BRASS.	ZINC.	TIN.
Golden yellow (refractory)	3	1	......
Half white (readily fusible)	12	5	1
White	20	1	4

In making these solders a good quality of sheet brass is used. Fuse the brass, then add the zinc and tin and stir thoroughly. Remove the crucible from the furnace as soon as the metals are thoroughly mixed.

Brass solders are commonly used in the granulated form. The granulation may be effected by pouring the melted alloy through a wet broom.

Silver Solders. Silver solders, which consist of silver alloyed with certain proportions of copper, zinc and often tin, have a broad application. They can be used in soldering articles of silver, brass,

German silver, cast iron, steel, etc. The small proportion of tin sometimes added renders the solder
more fusible and causes it to flow more easily.

The following formulæ are often recommended
for general use:

NO. 1.

Pure silver........................... 8 parts.
 "   copper........................, ... 1 "
 "   zinc............................. .... 2 "

Another solder may be made of :

NO. 2.

Pure silver........................ ...... 2 parts.
Brass wire .... 1 "

A hard silver solder for soldering articles which
are to be hammered or stamped, may be made as
follows:

NO. 3.

Silver .... 4 parts.*
Copper..................... 1   "

A solder for steel may be made of:

NO. 4.

Silver .... 3 parts.*
Copper 1   "

In preparing silver solder from the separate metals,
melt the silver under considerable borax, add the
copper and finally introduce the zinc as usual. In
case brass is used instead of copper and zinc sepa-
rately, add it after the silver has melted.

---

*Brannt's " Metallic Alloys."

Silver coin may be used instead of pure silver in preparing solders. The silver coin of the United States is composed of silver, ninety parts, and copper, ten parts.

Silver solder is commonly used in the form of clippings; hence, pour the metal into the upright ingot mold (Fig. 19) and roll the ingot, as described on page 111, into a plate of 26 or 28 gauge.

Gold Solders. These are alloys used in uniting articles composed of pure or alloyed gold. Besides gold, they contain copper, silver, and often some zinc. As several standards of gold are used by dentists and jewelers, solders of different degrees of fineness, usually ranging from 12 to 20 carats, are used.

Obviously, in order to obtain the most artistic, as well as the most substantial results, the color of the solder should correspond to that of the material upon which it is to be used, and its melting point should be but slightly lower. As stated above, zinc is sometimes added to gold solders. It reduces the melting point and improves the flow, but is objectionable, especially if in excess, owing to the fact that it renders the solder brittle and difficult to roll, and at times oxidizes to such an extent that it leaves the surface "pitted."

The following formulæ are suggested for making satisfactory solders for dental purposes. As gold coin is often employed in making gold solders, the proportions of this alloy, as well as those of pure gold to be used, are given.

NO. 1. 12-CARAT GOLD SOLDER.

Gold	12 parts.	Gold coin (U. S.)..	13.3 parts.
Silver.............	6 "	Silver	6 "
Brass.............	6 "	Brass.............	4.7 "

NO. 2. 13-CARAT GOLD SOLDER.

Gold	13 parts.	Gold coin (U. S.)..	14.4 parts.
Silver.............	6 "	Silver	6 "
Brass..	5 "	Brass.............	3.6 "

NO. 3. 15-CARAT GOLD SOLDER.

Gold	15 parts.	Gold coin (U. S.)..	16.6 parts.
Silver.............	4 "	Silver	4 "
Brass.............	5 "	Brass.............	3.4 "

NO. 4. 16-CARAT GOLD SOLDER.

Gold..............	6 parts.	Gold coin (U. S.)...	6.6 parts.
Silver............	2 "	Silver............	2 "
Brass.............	1 "	Brass.............	0.4 "

NO. 5. 18-CARAT GOLD SOLDER.

Gold..............	27 parts.	Gold coin (U. S.)....	30 parts.
Silver............	4 "	Silver............	4 "
Copper...........	2.5 "	Copper...........	0 "
Brass.............	2.5 "	Brass.............	2 "

The following solders are recommended for crown and bridge work :

NO. 6. 20-CARAT GOLD SOLDERS.*

Gold...	20 parts.
Silver...	2 "
Copper	1 "
Spelter solder (Cu1, Zn1).....................	1 "

LOW'S SOLDER.		RICHMOND'S SOLDER.	
Gold coin (U. S.)....	12 parts.	Gold coin (U. S..)....	5 parts.
Silver.............	2 "	Brass wire...........	1 "
Copper...........	1 "		

The method of preparing gold solder does not differ from that given for making silver and other

---

*Essig's "American Text-book of Prosthetic Dentistry."

solders. Only pure metals should be used; care should be taken to insure thorough mixing, and finally, the mold into which the solder is to be poured should be scrupulously clean. Gold solders should be cast in the upright ingot mold and rolled in the same manner as silver solders.

Solders for Aluminum Bronze. The difficulties encountered in soldering aluminum are not met with in soldering aluminum bronze. The following solders are recommended for ten per cent aluminum bronze:*

NO. 1. HARD SOLDER.		NO. 2. MEDIUM HARD SOLDER.	
Gold	88.88	Gold	54.40
Silver	4.68	Silver	27.60
Copper	6.44	Copper	18.00

NO. 3. SOFT SOLDER.

Bronze (Cu 70, Sn 30)	14.30
Gold	14.30
Silver	57.10
Copper	14.30

Platinum Solder. As already stated, platinum for chemical purposes is soldered autogenously. For other purposes, however, pure gold is used. Owing to the difficulty with which it is melted, soldering with pure gold requires great skill in order to obtain perfect joints.

Soldering.

In applying soft solders the soldering iron is commonly used. This consists of a pointed piece of copper called the *bit*, attached by an iron stem to

---

* Richards' "Aluminum."

a wooden handle. In order to solder successfully, the bit should be kept clean and tinned on the point. This can be done by heating it in a Bunsen burner, or preferably in a bed of charcoal, then rubbing in ammonium chloride (sal ammoniac) and finally in some soft solder.

In applying hard solders either a mouth or brazing blowpipe (Fig. 23) is used, as the highest temperature of the soldering iron is not sufficient to melt the solder.

In soldering, the following rules should be observed:

1. The surfaces to be united should be fitted together closely and, if necessary, held in position by binding-wire or in some other manner.

2. The surfaces should be scrupulously clean, i. e., free from oxides, grease, etc. This is often accomplished by scraping or polishing them, but more often by applying dilute acids as sulphuric or hydrochloric.

3. Some means should be adopted to prevent the air from coming in contact with the heated surfaces. This is especially necessary in hard soldering. For this purpose fluxes are used. In soft soldering the so-called *soldering fluids* are employed. They are designed to clean and to protect the surfaces and at the same time to cause the solder to flow. A satisfactory flux of this sort can be made by saturating commercial hydrochloric acid with zinc. Instead of this fluid common rosin can often be used. For hard soldering borax is the most common flux. It not only prevents oxidation but also removes oxides from the metals. It is best applied in the form of a thin paste made by mixing pulverized borax with water.

4. The quantity of both solder and flux should be the least that will suffice. An excess of either is very objectionable, especially in hard soldering.

5. In hard soldering the temperature should be raised gradually and care should be taken not to over-heat. When large surfaces are being soldered it is customary to carefully heat the entire piece uniformly before directing the flame upon the parts to be united.

6. An oxidizing flame should be avoided, as it oxidizes the base metals and thus interferes with the operation.

7. After soldering, the joint should not be dis-turbed until the solder has thoroughly hardened.

CHAPTER XVI.

MISCELLANEOUS ALLOYS.

In this chapter various alloys, some of which have important applications in dentistry, are considered. Since it would be impracticable for the student to attempt to prepare all of even the more important alloys described herein, it may be well to indicate that the preparation and study of *fusible alloys* offers an interesting and instructive laboratory exercise. Therefore, about twenty-five grams each of two of the more important representatives of this class, preferably Mellotte's and Wood's, may be made, and their physical properties, particularly their low melting point, noted.*

Fusible Alloys.

As already stated, the term *fusible alloy* is applied to a number of metallic compounds, composed chiefly of tin, lead, bismuth, and occasionally cadmium and mercury, which melt at very low temperatures, in some instances even below the temperature of boiling water. In most cases, especially if the proportion of bismuth is high, the alloy is brittle and considerably harder than any of the metals composing it.

*The form of report to be submitted upon the completion of these exercises is shown in the Appendix, Section II.

In the dental laboratory fusible alloys are used for crown and bridge work and for various other purposes. One of the most common alloys of this class is that introduced by Mellotte. It has the following composition : *

Bismuth	8 parts.
Tin	5 "
Lead	3 "

This alloy melts at about 100° C. and expands on solidifying. It is harder than tin, somewhat softer than zinc and quite brittle. In order to obtain impressions to be used with it, a compound of potter's clay and glycerine, called "moldine," is used. This substance, although retaining its plasticity for a long time, eventually becomes hard and has to be made plastic again by moistening with glycerine. Another fusible alloy, suggested by C. M. Richmond for dental uses, has the following composition :

Tin	20 parts.
Lead	19 "
Cadmium	13 "
Bismuth	48 "

This alloy is said to be as hard as zinc. It melts at about 65.5° C.

In addition to the alloys given above a table of others, some of which are often employed in dentistry, is given below.

---

*Essig's "American Text-book of Prosthetic Dentistry."

FUSIBLE ALLOYS.

Name.	Bismuth.	Tin.	Lead.	Cadmium.	Mercury.	Antimony.	Melting Point.
Hodgen's ...	8	3	5	......	......	2	105° C.
Darcet's	4	1	3	......	......	......	96° C.
Rose's.	2	1	1	......	......	......	95° C.
Newton's....	8	3	5	......	......	......	94° C.
Onion's.....	5	2	3	......	......	......	92° C.
Wood's.	4	1	2	1	......	......	65° C.
Lipowitz's ..	15	4	8	3	......	......	63° C.
Darcet's .... (with mercury)	2	1	1	......	10	......	45° C.

In preparing these alloys mix the constituents excepting the cadmium and mercury, and melt them under charcoal. As the tendency of lead to separate is very great, stir the alloy thoroughly with a pine stick before pouring. Cadmium and mercury should be added after the other metals are melted, otherwise they will be volatilized.

DETERMINATION OF MELTING POINT.

To determine the melting point of a fusible alloy, proceed as follows : Select a clean, thin-walled test tube and bind to it by means of rubber bands a thermometer in such a manner that the bulb of the thermometer lies against the bottom of the tube. Place the alloy to be tested in the tube and immerse the tube and its contents, with the thermometer attached, in a beaker of distilled water. The water is then heated, with constant agitation, over a hot-

plate until the alloy melts. The source of heat is withdrawn and the temperature at which the alloy *solidifies* is noted. The mean temperature of two or three of these observations is taken as the melting point of the alloy. Obviously water can be used only where the alloy fuses below 100° C. When the alloy to be tested melts at or slightly above this temperature some ammonium chloride or common salt may be added to the water to raise its boiling point, or a liquid possessing a higher boiling point than water, such as glycerine or sulphuric acid, may be used. For the determination of the melting point of an alloy which melts considerably above the tempera-
. ture of boiling water other methods must be employed which cannot be described here.

Alloys for Dies.

An alloy which is to be employed in making dies should possess the following properties: It should not be so brittle that it will break under the blows of a heavy hammer; it should shrink but slightly upon solidifying; it should fuse at a comparatively low temperature; and finally, it should flow easily in order to make a perfect reproduction of the model.

Pure zinc seems to meet most of the requirements called for above, and hence it is widely used for making dental dies. Although it contracts somewhat in the act of cooling, this is said by some to be immaterial since it serves to counteract the expansion which the plaster model undergoes in setting.

Crude zinc cannot be used for making dental dies

owing to the impurities in it which render it very brittle and greatly impair certain other physical properties. A variety which is highly recommended for dental uses is that known as Bertha zinc.*

Babbitt Metal. This alloy is made according to a great many formulæ. A compound of the following composition is recommended by Haskell for dental purposes:

Copper.........................	1 part.
Antimony........................	2 "
Tin.............................	8 "

This alloy is said to be nearly as hard as zinc, and to contract less. Common Babbitt will not do for dies.

In preparing Babbitt metal, melt the copper and half of the tin, then add the antimony and the remainder of the tin. Stir vigorously and keep the surface of the alloy covered with powdered charcoal. Babbitt metal deteriorates when melted repeatedly.

Spence's "Metal." This is a compound occasionally used in making dies. It is not strictly an alloy, but a solution of the sulphides of metals, as lead, iron, antimony, zinc, etc., in melted sulphur. For dental purposes the compound made by melting iron sulphide in sulphur is commonly used. It melts at 160° C. and expands on cooling. It gives very good castings.

Type Metal. Type metal, which is used in the manufacture of type and frequently in dentistry for making dies, varies considerably in composition. A good variety is composed of:

---

*Essig's "American Text-book of Prosthetic Dentistry."

```
Lead............................. 4 parts.
Antimony......................... 1   "
Tin.............................. 1   "
```

It is brittle, harder than lead, softer and more fusible than zinc, and is capable of giving accurate castings.

In making type metal melt the lead first, then add the tin, and finally the antimony. Carefully regulate the temperature of the furnace in order not to oxidize the antimony. Keep the alloy covered with charcoal.

Alloys of Zinc and. Tin. An alloy suggested by Richardson for making dies and described as being harder and more fusible than zinc is composed of zinc, four parts, and tin, one part. Similar alloys are frequently employed in casting patterns and ornaments.

Alloy for Counterdies.

The alloy commonly employed for counterdies with Babbitt metal dies consists of lead, seven parts, and tin, one part.

Gold Plate.

For dental purposes gold plate of 18 or 20 carats is commonly used. A lower carat than 18 is not desirable since it discolors readily and is more or less acted upon by the saliva ; a higher carat than 20 lacks rigidity to withstand the strain of mastication when used for artificial dentures. When it is required, however, to employ higher grades of gold plate they are usually alloyed with small quantities of platinum, and this renders them very rigid. The formulæ of several varieties of gold plate are given below.

18-CARAT GOLD PLATE.

Pure gold...........18 parts.	Gold coin (U. S.).... 20 parts.
Copper.............. 2 "	Copper .... 0 "
Silver .... 4 "	Silver .... 4 "

20-CARAT GOLD PLATE.

Pure gold.......... 20 parts.	Gold coin (U. S.).....22 parts.
Copper.............1.5 "	Copper........... 0 "
Silver . ...2.5 "	Silver............. 2 "

22-CARAT GOLD PLATE.

Pure gold	22 parts.
Copper...........................	0.25 "
Silver...........................	1.50 "
Platinum........................	0.25 "

18-CARAT GOLD PLATE FOR CLASPS, HARD WIRE, ETC.

Pure gold...........18 parts.	Gold coin (U. S.).....20 parts.
Copper............ 3 "	Copper........... 1 "
Silver............ 2.25 "	Silver............. 2.25 "
Platinum......... 0.75 "	Platinum.......... 0.75 "

In the preparation of gold plate no directions beyond those already outlined for the preparation of gold solders need be given. The alloy should be poured into the clean, warm, upright ingot mold. After annealing it is rolled into a plate of any required gauge, as directed on page 111.

TABLE OF MISCELLANEOUS ALLOYS.

Name.	Gold.	Silver.	Copper.	Tin.	Lead.	Zinc.	Nickel.	Aluminum	Antimony.
Gold coins—									
United States...	900	....	100	....	....	....	....	....	....
English........	917	....	83	....	....	....	....	....	....
Silver coins—									
United States...	....	900	100	....	....	....	...	....	....
English.......	....	925	75	....	....	....	....	....	....
Mexico (*peso*) ..	....	901	99	....	....	....	....	....	....
Alloys resembling gold—									
Dutch gold.....	....	....	84.5	....	....	15.5	....	....	....
Nürnberg gold.	2.5	....	90	....	....	....	....	7.5	....
Green gold.......	75	25	....	....	....	....	....	....	....
Red gold........	75	....	25	....	....	....	....	....	....
Bronzes—									
For statues....	....	....	91	2	1	6	....	....	....
Gun metal....	....	....	90	10	....	....	....	....	....
Bell metal.....	....	....	80	20	....	....	....	....	....
Alumin. bronze.	....	..	90	....	....	....	....	10	....
German silver....	....	....	60	....	....	20	20	....	....
Brass...........	....	....	66.6	....	....	33.3	....	....	....
Britannia metal (English).......	....	....	1.84	81.90	....	....	....	....	16.25
Hercules metal...	....	....	*88	....	10	2	....	2.5	....
Pewter...........	....	....	....	80	20	....	....	....	....

*Bronze.

♦

CHAPTER XVII.

DENTAL CEMENTS.

Three varieties of cement are used in dentistry under the names of *oxyphosphate, oxychloride* and *oxysulphate* of zinc. The materials used in preparing these cements consist of a powder which is chiefly zinc oxide and a liquid which in the case of the first is a solution of glacial phosphoric acid, of the second, a saturated solution of zinc chloride and of the last named, a dilute solution of zinc chloride or of gum arabic. When the powder and the liquid are mixed on a glass or on a porcelain plate with a spatula a mass is produced which in time owing to some inexplicable change assumes a hard state. The use of oxychloride and oxysulphate cements is quite limited at the present time. Oxyphosphate, however, is extensively employed as a filling material and for various other purposes.

The chief objection to all cements, particularly to the oxychloride and oxysulphate, is that they are readily acted upon by the fluids of the mouth and are therefore of little value as permanent fillings. This tendency to disintegrate is probably facilitated somewhat by the change of volume, usually contraction, which they undergo in hardening and by their more or less porous character. For the purpose of counteracting the solubility and of conferring greater hardness upon the mass when "set" small quantities of such substances as silica, ground glass, borax, etc., are added. The value, however, of these constituents is questionable. Although it is impossible to render cements insoluble, it is a noticeable fact that their permanence is dependent to a great extent upon the manner of preparing, keeping and mixing them. In all cases the materials must be properly prepared. It is a fact to be observed that the setting and other properties of cements are dependent to a high degree upon the dehydrated condition of the powder. Hence, in working cements care should be taken to protect the powder from moisture. Finally, it is observed that to give the best results a cement must be thoroughly and carefully spatulated into a homogeneous mass.

Since it is highly essential that the student have some understanding of this important class of substances, an outline is given below to be followed in preparing, testing and otherwise studying them.*

Oxyphosphate Cement.

As already stated, oxyphosphate has almost entirely replaced the other cements for most purposes. In this cement the powder usually is specially prepared zinc oxide. Some manufacturers, however, add beside the substances already enumerated small

---

* The form of report to be submitted in this work is shown in the Appendix, Section II.

proportions of tin or of bismuth oxide.*   It is sometimes stated that a satisfactory cement powder can
be made simply by calcining the crude zinc oxide. It
will be found, however, that the cement will prove more
satisfactory if the zinc oxide is treated with nitric
acid previous to calcining.   This treatment gives it a
characteristic "dry fineness," and renders it somewhat gritty, no matter how carefully it may have
been pulverized and sifted.

In the coloring of oxyphosphate cement, and this
applies as well to the other varieties, it is seldom necessary to use any coloring matter.   By properly
manipulating the heat, colors from a light cream to a
dark yellow can be obtained. To make a light cement
a heat somewhat above redness may be employed.
The common dark yellow is best prepared by applying a white heat for about two hours.   In certain
cases, however, when special colors are required,
yellow ocher, lampblack, ferric oxide, etc., may be
added.

It is generally recognized that oxyphosphate
cement is more or less irritating when placed in
proximity to the pulp.   A suggestion of the possible
cause of this action is found in the fact that cement
powders contain more or less arsenious oxide.   The
reason for this is that zinc and arsenic are often
closely associated in nature, and when zinc oxide is
prepared for a cement powder the treatment often
fails to remove all the arsenic.   Out of fifteen promi-

*For the composition of several cements as determined by
analysis, see Appendix, Section II.

nent cements analyzed by the author, but two were found to contain only traces, while the remainder contained from 0.04 to 0.1 per cent of arsenious oxide.* Although it is impossible by ordinary methods entirely to eliminate arsenic from a cement, the quantity is reduced to a mere trace by the treatment already described.

A common complaint with oxyphosphate cement is the tendency of the liquid to change with time. This change sometimes shows itself in the formation of small crystals. This can be corrected by carefully adding one or more drops of water. In some cases, however, the liquid becomes turbid, and finally forms a paste-like mass. Of course, in this condition, the cement will work imperfectly. The causes of some of the changes which cement liquids undergo as the result of "ageing" are not definitely known. It may be stated, however, that the purity of the acid employed and the care bestowed in preparation are controlling factors in these changes.

Instead of the solution of glacial phosphoric acid, the crystals are sometimes used. They are carefully melted in a spoon, preferably of platinum, without boiling, and after the liquid has cooled somewhat it is mixed as usual with the powder.

PREPARATION OF OXYPHOSPHATE CEMENT.

Powder. Weigh out in a porcelain dish on the laboratory scales about forty-five grams of crude zinc oxide. Moisten with concentrated nitric acid, place

---

*See table of analyses of cements in the Appendix, Section II.

over a Bunsen burner and apply a gentle heat, continually stirring with a glass rod until brown fumes
cease to come off. Next transfer the powder to a
clean clay crucible and place it in the furnace. Apply
a white heat for one or two hours, then remove from
the furnace, pulverize in a mortar, sift through bolting cloth and bottle at once. Take of this fifteen grams
in a separate bottle, and retain the remainder for use
later in making the other cements.

Liquid. For the liquid take ten cubic centimeters
of distilled water in a test tube and add several pieces
of glacial phosphoric acid. Allow this to stand, applying gentle heat at times and keeping plenty of the
acid in the tube. When the liquid has reached about
the consistency of glycerine and is found to mix sat-
isfactorily with the powder, it may be filtered through
glass wool to remove any suspended matter and
then bottled.

MIXING OXYPHOSPHATE CEMENT.

In mixing remove the required quantity of liquid
to the mixing plate by means of a spatula, then pour
out the powder about an inch from the liquid. *Do
not introduce the spatula into the bottle containing the
powder.* Add small quantities of the powder to the
liquid, spatulating each portion thoroughly with a
broad, stiff-bladed spatula until the required consistency is reached. Since the success of a cement
depends largely upon the manner in which it is
mixed, the student should not fail to spatulate thoroughly until a homogeneous mass is obtained.

In preparing the samples of cement for the tests which follow, the powder should be added until a putty-like mass is produced.

TESTING OXYPHOSPHATE CEMENT.

1. General Tests. The following tests* have been suggested as characterizing a good oxyphosphate cement.

1. When first mixed it should yield a tough mass which when removed from the spatula does not adhere to the fingers and can be rolled into a pliable pellet.

2. It should have a glassy surface, and at the end of two or three minutes it should rebound when dropped upon wood, glass or porcelain.

3. At the end of five minutes it should be quite hard and should sound like porcelain when tapped.

4. After ten or fifteen minutes it should be dented with difficulty, and when broken should show a clean, sharp fracture.

5. After twenty minutes it should be very hard and should be capable of taking a good burnish.

6. In thirty minutes it should have little or no acid taste.

Poor cements will often become granular when first mixed, and will not so harden but that they can be cut like plaster of Paris after an indefinite time. Moreover, they will often fail to rebound, and will usually have a sticky feeling when pressed between the fingers. Finally, they will often retain for hours,

---

*See Flagg's '' Plastics and Plastic Filling.''

or even for days, an astringent taste. No rule can be
given for correcting these difficulties. The cement
powder should be recalcined, and if still unsatis-
factory the liquid may be thickened with a little more
phosphoric acid, or made somewhat thinner by the
addition of a drop or two of water. It will be necessary for the student to experiment until the desired
results are attained.

2. *Tests for Change of Volume.* In making these tests
a sufficient quantity of cement should be mixed to fill
a cavity in one of the steel blocks (Fig. 29) used in
testing amalgams. The expansion or contraction
can then be measured by means of the micrometer
(Fig. 28). Microscopical examinations of the margins should accompany the micrometrical measurements.

3. *Tests for Strength.* Some idea of the strength
of cements can be gained by preparing blocks as
described in the corresponding tests under amalgams
and by crushing these samples in the dynamometer.
Better results will be obtained if a matrix furnishing
larger blocks than that described is used. A good
sized block is 0.125 x 0.125 x 0.125 of an inch. These
samples, of course, should be given plenty of time to
harden before they are crushed.

4. *Tests for Porosity.* Cements are more or less
porous, as can be shown by the following tests: Mix
some cement and roll it into several pellets of equal
size. Drop these into some red ink or other colored
liquid in a test tube. Remove the pellets successively
at intervals of one or two hours, break them open

and note how far the coloring matter has penetrated them. Some cements will be found to color throughout in a short time, while others will require a much longer time.

5. Tests for Arsenic. In testing for arsenic in cement powders, the Marsh test (see pages 24 and 114) is used. About five grams of the material are dissolved in the smallest possible quantity of moderately concentrated hydrochloric acid. When the evolution of hydrogen in the apparatus has continued for some time and has been tested and found free of arsenic by holding a piece of cold porcelain in contact with the flame, the solution to be tested is poured in through the funnel tube, and a piece of porcelain is again held in contact with the flame. If the cement in question contains arsenic, a more or less heavy deposit of metallic arsenic will be obtained. Since arsenic is readily volatilized by heat, the deposit will be more marked if the porcelain is not held in one position, but is moved about so as to keep a cool surface against the flame.

The test just described is used in preference to that employed in qualitative analysis, since it is capable of detecting smaller quantities.

After applying the above tests to the cement prepared as directed, the student should in the same manner examine some well-known oxyphosphate cements found upon the market.

Oxychloride Cement.

Oxychloride cement is used chiefly for lining cavities prior to filling them. It is said to exert a preservative action toward teeth, but is more or less irritating. This is commonly attributed to the zinc chloride solution. As already stated, it shrinks considerably, is readily acted upon by the fluids of the mouth and hence is valueless as a material for permanent fillings.

PREPARATION OF OXYCHLORIDE CEMENT.

Powder. In preparing the powder for this cement proceed as follows :   Weigh out on the pulp balance (Fig. 4) ten grams of the calcined zinc oxide already prepared. Thoroughly mix with this one-tenth of a gram of borax and two-tenths of a gram of silica. Transfer to a clay crucible and calcine in the furnace for one-half hour or so at a bright red heat. Remove, grind in a mortar, sift through bolting cloth and bottle.

Liquid. To prepare the liquid take 10 c. c. of pure concentrated hydrochloric acid and add metallic zinc, applying gentle heat occasionally until the acid is saturated. Finally, filter through glass wool to remove any black specks or pieces of undissolved zinc and transfer to a tightly stoppered bottle.

MIXING OXYCHLORIDE CEMENT.

The directions given for mixing this cement do not differ from those given in mixing oxyphosphate cements. A good oxychloride will set in fifteen to twenty minutes, but requires several hours in which to reach its hardest state.

Oxysulphate Cement.

This cement is usually deficient in hardness but as generally stated being nonirritating, it serves a useful purpose for protecting pulps and for various other purposes.

PREPARATION OF OXYSULPHATE CEMENT.

Powder. Take ten grams of the zinc oxide already prepared and mix with it four grams of dry zinc sulphate. Transfer to a crucible and calcine in the furnace as directed in the preparation of oxychloride. Pulverize, sift and bottle.

Liquid. Dissolve two grams of zinc chloride in ten cubic centimeters of water. This yields a turbid liquid which should be shaken when used.

MIXING OXYSULPHATE CEMENT.

The directions given for mixing this cement differ but slightly from those just outlined for oxychloride. The powder is added to the liquid until a mass not thicker than cream is obtained. This cement is ready for use when it shows the slightest tendency to thicken after being worked for some time with the spatula.

CHAPTER XVIII.

SPECIAL PROBLEMS.*

The Analysis of Teeth.

Roughly speaking, the solid matter of a tooth is composed of thirty per cent of organic and seventy per cent of inorganic matter. The organic matter is chiefly a substance resembling the ossein in bone. Like ossein, it is more or less completely converted into gelatine by prolonged boiling with water. The inorganic or earthy matter consists mostly of calcium phosphate with some magnesium phosphate, calcium carbonate and small quantities of other salts. The chemical composition of a tooth, however, is not the same throughout. Certain parts, namely, the dentine, the bulk of the tooth and the cement which covers the roots, resembles the bone in composition. On the other hand, the enamel which invests the crown of the tooth contains but little organic matter and is composed chiefly of calcium phosphate and fluoride.

In the table given below the composition of the different parts of the tooth is compared. These results fail to take into account the water, which constitutes about five or six per cent of the total weight of the tooth.

---

*The forms of the reports to be submitted after performing the exercises in this chapter are shown in Appendix, Section II.

COMPOSITION OF DENTINE, CEMENT AND ENAMEL COMPARED.

(VON BIBRA.)

	Dentine.	Cement.	Enamel.
Inorganic matter	71.99	70.58	96.41
Organic matter	28.01	29.42	3.59
	100.00	100.00	100.00
Calcium phosphate and fluoride	66.72	60.73	89.82
Calcium carbonate	3.36	8.02	4.37
Magnesium phosphate	1.18	1.00	1.34
Other salts	0.73	0.83	0.88
Organic matter	27.61	28.70	3.39
Fat	0.40	0.72	0.20
	100.00	100.00	100.00

The inorganic constituents of teeth are insoluble in water, but are readily soluble in dilute acids by the action of which they may be removed, leaving the pliable organic matter which retains the general form of the tooth.

The simple exercises given below will serve to show some of the facts just stated. In these tests no attempt will be made to show the differences in composition between the dentine, the cement and the enamel, but the tooth will be considered as a whole.

QUANTITATIVE TESTS.

1. Estimation of Water. Crush in a mortar a freshly extracted tooth which has been freed of adhering matter.* Weigh out one gram on the analytical balance in a porcelain crucible, the weight of which is known. Place the crucible and its contents in an air bath for one hour, at a temperature of 95° C. At the end of this time remove the crucible, allow to cool and weigh as quickly as possible, in order that the hygroscopic residue be prevented from absorbing moisture from the air. Return the crucible, after weighing, to the bath, and allow it to remain there one-half hour. Again cool and weigh. In case the weight obtained here corresponds with the previous weight the sample may be considered dehydrated. When, however, the second heating in the bath reduces the weight it will be necessary to again heat until a constant weight is obtained. The loss in weight, i. e., the difference in weight between the original and that obtained after drying, equals the weight of water in the sample. This multiplied by 100 gives the percentage.

2. *Estimation of Organic and Inorganic Matter.* Place the crucible containing the dehydrated sample on the pipestem triangle. Cover with a crucible cover and ignite gently at first until the organic matter is burned. Remove the cover and calcine at a red heat until the ash is nearly white. Cool and weigh. The

*In case a freshly extracted tooth cannot be obtained, one which has been extracted for some time can be used. Before using it for these tests, however, it should be placed in water for twenty-four hours or so, in order that the results for water may be comparable with those which would be obtained if a fresh tooth were used.

loss in weight over that occasioned by the dehydration represents the weight of organic matter in the
sample. This is converted into percentage by multiplying by 100. To obtain the percentage of inorganic
matter add the percentage values obtained for
organic matter and water and subtract this result from
100.

QUALITATIVE TESTS.

1. For Organic Matter. Place a tooth in dilute
hydrochloric acid in a test tube. After two or three
days the inorganic matter will be removed and the
soft, elastic organic matter will remain. Boil this
with water to convert into gelatine, which solidifies
to a jelly on cooling.

2. For Phosphates. Transfer the inorganic residue left from the *Quantitative Tests* given above to a
test tube, and dissolve it with heat in dilute hydrochloric acid. Filter out any insoluble specks and
reserve the filtrate for the tests which are to follow.
To a few drops of this solution add a drop or two of
nitric acid, then an excess of ammonium molybdate,
and boil. Phosphates are indicated by the yellow
crystalline precipitate of ammonium phospho-molybdate.

3. For Calcium and Magnesium Phosphates. To
another portion of the hydrochloric acid solution
add ammonium hydroxide to alkaline reaction. A
bulky white precipitate of calcium and magnesium
phosphates appears. Filter and retain the filtrate for
the test given below.

4. For Calcium. To the clear filtrate just obtained add ammonium oxalate. A white precipitate of calcium oxalate appears, showing that some calcium is present in the tooth uncombined with phosphoric acid. This is the calcium present as calcium carbonate, further indicated by the slight effervescence when the hydrochloric acid was added to the inorganic residue.

ACTION OF VARIOUS SUBSTANCES ON THE TEETH.

It is evident from facts already observed that the inorganic matter of teeth is readily dissolved by acid, and it follows then that tooth structure can be greatly injured by coming in constant contact with substances of an acid character. Among the common mineral acids which may exert a disintegrating action upon teeth are hydrochloric, nitric, sulphuric and phosphoric, and among the vegetable acids are acetic, tartaric, citric, lactic, etc. Furthermore many other substances, such as ferric chloride (tincture of iron), alum, potassium tartrate (cream of tartar), compounds of mercury, solutions of hydrogen dioxide, etc., produce injurious effects if brought constantly in contact with the teeth.

Analysis of Urine.

Much research has been made in urine analysis with the result that the composition of urine in health and in disease is quite well known. In health it always contains certain substances in quite constant proportions. Many grave disorders, however, are accompanied by wide variations in the proportions of the normal constituents and by the appearance of certain foreign substances.

In the table given below the average composition of urine passed during twenty-four hours is shown. Although the acids and metals are separated in this table, it must not be inferred that they exist so in the urine. They are found in the form of salts : Urates, sulphates, phosphates, chlorides, etc.

THE COMPOSITION OF NORMAL URINE.

(Twenty-four hour sample.)

Water	1500.00	grams.
Total solids	72.00	"
Urea	33.18	grams.
Uric acid	0.55	"
Hippuric acid	0.40	"
Creatinine	0.91	"
Pigments and other organic matter..	10.00	"
Sulphuric acid	2.01	"
Phosphoric acid	3.16	"
Chlorine	7.50	"
Ammonia	0.77	"
Potassium	2.50	"
Sodium	11.09	"
Calcium	0.26	"
Magnesium	0.21	"

It would be impracticable in this course to attempt a complete chemical analysis of urine. Indeed, such is unnecessary in the analysis of urine for clinical purposes. In the following outline only such tests and processes will be given as are needed to meet the requirements of practical work. For more details concerning the significance of certain abnormalities the student is referred to books upon the subject.

OUTLINE OF TESTS.

1. Observation of odor, color, reactions, etc.
2. Determination of specific gravity.
3. Tests for albumin.
4. Test for mucin.
5. Tests for sugar.
6. Tests for abnormal coloring matter.
7. Tests for phosphates and chlorides.
8. Determination of uric acid.
9. Determination of urea.

1. PHYSICAL CHARACTERISTICS.

Normal urine is of a pale yellow or straw color, shading at times to a reddish yellow. If blood is present urine may have a reddish brown tint, or greenish brown, if bile is present. Fevers, certain drugs, and an increase or decrease in the quantity of liquids consumed will produce a marked effect upon the color of urine.

At different times during twenty-four hours the urine may be alkaline. The whole mixed product, however, should be slightly acid in reaction, due, it is said, to the presence of acid phosphates, to uric acid and to certain other organic acids. In taking the reaction, use both the red and the blue litmus paper.

Fresh urine is clear and possesses a peculiar aromatic odor. Any turbidity is usually due to precipitated phosphates or urates. If phosphates, they will dissolve with a drop of acid; if urates, they will redissolve with heat. Pus yields a deposit similar to that of phosphates. It may be detected under the microscope, or by the ropy, gelatinous mass formed by the addition of potassium hydroxide to the deposit formed after standing.

2. DETERMINATION OF SPECIFIC GRAVITY.

The quantity of urine passed by a healthy individual in twenty four hours may be taken as about 1500 c. c. and its specific gravity will range from 1.015 to 1.025; 1.020 is usually taken as the normal.

Owing to the fact that in disease the specific gravity may run much below or above the normal value, its determination is of considerable importance in the examination of urine for clinical purposes. To be of value, however, the specific gravity should always be taken upon a sample of the entire quantity excreted in twenty-four hours. For a very exact determination a special balance known as the Mohr-Westphal should be used, although in practical work a good *urinometer* (Fig. 36) is considered sufficiently accurate. In using the urinometer pour the accompanying cylinder nearly full of urine, allow the air bubbles to escape, bring to a proper temperature and immerse the urinometer in it. Read the specific gravity indicated on the stem below the surface of the urine.

Wide variations in specific gravity nearly always

occur in certain diseases, particularly in *Bright's disease* and in *diabetes mellitus.* In the former the specific gravity is usually low on the twenty-four hour sample, ranging from 1.015 down. On the other hand, *diabetes mellitus* is accompanied by a great increase in specific gravity, often reaching 1.040 or higher. Quite wide variations in specific gravity are often observed in health. These variations result from perfectly natural causes. For example, the consumption of large quantities of liquids will have a tendency to decrease the specific gravity, while on the other hand excessive perspiration or the consumption of small quantities of liquids and large quantities of foods will greatly increase the specific gravity, often to 1.030, or higher.

3. TESTS FOR ALBUMIN.

Albumin is not a constituent of normal urine, although it may sometimes be found in the urine of healthy individuals. In such cases it is often due to slight disorders not difficult to regulate. Albumin may appear persistently in the urine, however, in quite large quantities as the result of the grave disorder known as Bright's disease. Of the various tests proposed for detecting albumin the following are usually satisfactory in practical work. *Before applying the following tests the urine should be filtered if cloudy.*

1. Coagulation by Heat. To a test tube nearly full of urine add ten or twelve drops of strong nitric or acetic acid. Boil the upper part of the tube. A cloud indicates albumin; heat, alone, has the power of coagulating albumin, as is shown in the boiling of the white of egg. The purpose of the acid here is to prevent the precipitation of earthy phosphates which usually occurs when urine is boiled.

2. Coagulation by Nitric Acid. Heat a little strong nitric acid in a test tube, and by means of a small pipette carefully drop an equal volume of urine on the acid so as to perfectly overlie it without mixing. If albumin is present a whitish ring of coagulum will appear at the junction of the two liquids. The test just given may be made with cold nitric acid. In this case, however, the acid should be brought under the urine by means of the pipette. This test is not as reliable as the first, because at times an excess of urea in the urine may cause a ring to appear when albumin is absent. Heating the acid prevents this.

4. TEST FOR MUCIN.

In small quantities mucin has no pathological significance, since it is a constituent of normal urine. In catarrh of the urinary tract it may be present in considerable quantity, and in such cases it will cause a turbidity in the urine when first passed, or soon after, and will partly settle and float near the bottom of the vessel. Since the turbidity caused by mucin does not clear up with acetic or dilute nitric acid, it may be mistaken for albumin. They may be distinguished as follows: Half fill a test tube with acetic acid, and by means of a pipette add the urine so that they do not mix. Mucin is indicated by a cloud above the junction of the two liquids, while albumin appears at the junction.

5. TESTS FOR SUGAR.

While minute traces of sugar* (dextrose) are said to occur in normal urine, relatively large quantities indicate the pathological condition known as diabetes mellitus. Urine containing sugar possesses a high specific gravity, 1.030 or higher, and the quantity is greatly increased.

While many tests are proposed for detecting sugar, the following are reliable if properly carried out.

1. Fehling's† Test. Take 5 or 10 c. c. of Fehling's solution in a test tube and boil. No precipitation should take place. Next add the clear urine drop by drop, boiling a moment after each addition. If no precipitate forms add a volume of urine equal to that of the solution and again boil a moment. If sugar is present in even small quantities a yellowish precipitate will be formed, which, upon considerable boiling, changes to red. (Cu_2O.) The above test is used quantitatively as well as qualitatively, but involves more detail than can be given here. For ordinary purposes, a rough estimation of the quantity of sugar present can be made by noting the number of drops necessary to bring about a reduction. A precipitate occurring after adding but one or two drops of urine would indicate a large percentage of sugar.

---

*The student must not confuse the sugar found in urine, commonly known as dextrose or grape sugar, with common cane sugar. The latter does not reduce Fehling's solution.

†Directions for making and keeping Fehling's solution are given in the Appendix, Section I.

At times an excessive quantity of uric acid and certain other substances may produce an imperfect reduction of the Fehling's solution, shown by the change from the blue to a greenish color, or to a nearly colorless solution. This, of course, is not to be taken as evidence of sugar. In cases where such difficulties are encountered the following test may be used:

2. *The Bismuth Test.* To some urine in a test tube add an equal volume of ten per cent solution of potassium hydroxide and then as much pure bismuth subnitrate as can be held on the point of a penknife. Boil. If sugar is present a black substance consisting chiefly of lower oxides of bismuth is formed. If albumin is present in the urine it should be coagulated and removed by filtration before applying this test.

6. TESTS FOR ABNORMAL COLORING MATTER.

In disease coloring matter, such as bile and blood, may appear in the urine. Bile occurs in jaundice and gives the urine a dark brown, greenish, or, at times, a black color. Blood may appear as the result of hæmorrhage in any part of the urinary tract, and if present in large quantities will impart to the urine a deep red color. The following simple tests may be employed in detecting the presence of these substances:

1. For Bile Pigments. Pour a small quantity of urine into a test tube and introduce below it by means of a pipette an equal volume of strong nitric acid containing some nitrous acid. If bile is present a series of colors, the most characteristic of which is green, will be seen at the junction of the two liquids. (Gmelin's test.)

Another test suggested for detecting bile pigments is as follows :   Add a few drops of tincture of iodine to some urine in the test tube in such a manner that the iodine solution floats on the surface.   A green color at the junction indicates bile (Trousseau's test).

2. For Blood and Blood Pigments. Examine a small quantity of urine by boiling with potassium hydroxide.   The precipitated earthy phosphates will be colored red if the blood is present.

Another test for blood consists in examining the urine under the microscope and detecting the corpuscles which may be present.

7.   TESTS FOR PHOSPHATES AND CHLORIDES.

Both exist normally in the urine, although in disease they may be increased or diminished in quantity. To be of value for clinical purposes quantitative tests must be made.   Here, however, only qualitative tests will be given to prove their presence.

1. For Phosphates. The earthy phosphates usually precipitate with heat alone.   To remove these add ammonium hydroxide and boil.   Still another form exists in urine, namely, the alkali phosphates.   To detect them remove the precipitate of earthy phosphates by filtration and add *magnesia* mixture.*   This precipitates in the form of ammonium magnesium

---

* A solution containing magnesium sulphate, ammonium chloride and ammonium hydroxide.

phosphate the phosphoric acid which was combined with the alkali metals.

2. *For Chlorides.* The chlorides in urine exist chiefly in the form of sodium chloride or common salt. To detect them acidify a small portion of urine with nitric acid and add silver nitrate. A white precipitate of silver chloride appears.

8. THE DETERMINATION OF URIC ACID.

Uric acid occurs normally in the urine in quantities varying from 0.4 to 0.8 of a gram daily. It may be detected often in the form of reddish crystals in the bottom of a vessel in which urine has stood for some time. The clinical significance of variations in the quantity of uric acid is not very well known.

To merely detect uric acid add 10 c. c. of strong hydrochloric acid to 100 c. c. of urine and allow the liquid to stand in a cool place for one or two days. Next, filter and wash with water. Transfer to a porcelain dish, add a drop of strong nitric acid and evaporate to dryness on a water bath. The brown residue will turn purple with a drop of ammonium hydroxide (Murexid test).

A rough method for determining uric acid quantitatively is as follows: Add 25 c. c. of strong hydrochloric acid to 250 c. c. of urine and allow to stand as directed above. Collect the sediment on a weighed filter and wash with a little cold distilled water. Dry and weigh. The result is only approximate. In case albumin is present it should be removed before this test is made.

9. THE DETERMINATION OF UREA.

Urea is the most abundant solid constituent in urine. In health the average amount excreted is thirty-three grams or about two per cent, although it may vary widely under different conditions since it is dependent upon the quantity and quality of food consumed, and upon the losses by perspiration, etc. An increase in the quantity of urea accompanies diabetes mellitus, fevers, etc., and a decrease is seen in diseases of the liver and kidneys.

To prove merely the presence of urea in urine concentrate 50 c. c. to one-quarter its original volume and add some strong nitric acid. The urea nitrate formed separates as glistening crystals. This test is, of course, of no clinical importance, since all urine contains urea.

A method commonly employed for determining urea depends upon the fact that a solution of sodium hypobromite decomposes this substance into carbonic acid, nitrogen and water.

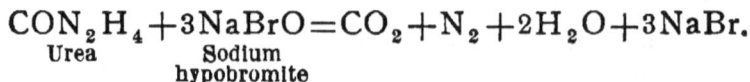

$$\underset{\text{Urea}}{CON_2H_4} + \underset{\substack{\text{Sodium} \\ \text{hypobromite}}}{3NaBrO} = CO_2 + N_2 + 2H_2O + 3NaBr.$$

If now a strongly alkaline solution of sodium hypobromite* is used the carbon dioxide is absorbed and from the volume of nitrogen set free may be calculated the weight of the urea present in the volume of urine employed. The apparatus commonly employed is illustrated in Fig. 37. Accom-

---

*For the preparation of this solution see Appendix, Section I.

panying the illustration is a description of the apparatus and the manner of using it.

Analysis of Saliva.*

The saliva is secreted by three pairs of salivary glands, the parotid, the submaxillary and the sublingual. The secretions from these glands differ somewhat in composition, but are mixed in the mouth, and to them are added salivary corpuscles, epithelial cells, etc. In the table given below the average composition of human saliva is given. It will be observed that it consists of about 99.5 per cent of water and 0.5 per cent solids.

THE COMPOSITION OF MIXED SALIVA (Jacubowitsch).

Water	99.51
Solids	0.49

Soluble organic bodies, ptyalin, etc	0.130
Epithelium	0.160
Inorganic salts	0.182
Potassium sulphocyanate	0.006
Potassium and sodium chloride	0.084

In the following outline of analysis only the more important tests and processes† will be given, and for more facts concerning the significance of abnormalities the student should look to some other

---

*To collect saliva for this work, wash out the mouth thoroughly with water, then chew a piece of rubber to excite secretion.

†A microscopical study of the sediments in both urine and saliva is necessary to render the analysis complete. This subject cannot, however, be touched upon in this book.

source. As will be seen, the general method of procedure differs but slightly from that outlined for the analysis of urine.

1. Observation of color, odor, consistency, reaction, etc.
2. Determination of specific gravity.
3. Tests for albumin.
4. Test for mucin.
5. Test for ptyalin.
6. Test for sulphocyanates.
7. Tests for chlorides, phosphates and sulphates.
8. Tests for calcium and magnesium.
9. Test for mercury.

1. PHYSICAL CHARACTERISTICS.

Normal saliva is colorless, odorless, viscid, frothy and slightly turbid. Conditions said to give rise to odor in saliva are gingivitis, scurvy, mercurial salivation, etc. In reaction saliva is alkaline, due it is said to the presence of alkali bicarbonates and phosphates, but the degree of alkalinity varies, being greatest during and after meals. Conditions which may render the saliva acid in reaction are the decomposition of organic matter in the mouth, diabetes mellitus, acute rheumatism, mercurial salivation, etc. Since it is often difficult to determine the reaction, especially when nearly neutral, both red and blue limus paper should be used.

2. DETERMINATION OF SPECIFIC GRAVITY.

The average amount of saliva secreted in twenty-four hours is about 800 or 900 c. c. Obviously it is impossible to determine the exact quantity since it depends upon so many conditions. The sight or even the thought of foods, the mastication of dry foods, the filling of teeth, certain drugs and certain diseases may greatly increase the quantity of saliva. On the other hand, fevers and inflammatory diseases decrease the quantity of saliva in a marked degree. Specific gravity of normal saliva may range from 1.002 to 1.009. In taking the specific gravity use the urinometer as directed under urine analysis. In case the quantity of saliva is insufficient to fill the cylinder, measure out a certain quantity of it and dilute it with an equal volume of distilled water and multiply the last figure of the specific gravity obtained under these conditions by 2. The result is the specific gravity of the sample. '

3. TESTS FOR ALBUMIN.

In testing for albumin, which is a normal constituent of saliva, apply the tests outlined under urine analysis. Before doing so, however, dilute and filter the portion to be tested.

4. TEST FOR MUCIN.

Mucin is an important constitutent of normal saliva and may be detected as outlined under urine analysis.

5. TESTS FOR PTYALIN.

Ptyalin is a diastatic ferment in saliva which has the power of converting starch into dextrin and maltose, the latter being a form of sugar which has the power of reducing Fehling's solution. Upon these facts is based a method for detecting it. The following tests will serve to prove the presence of ptyalin and to show its digestive power : Make a little starch paste by mixing one gram of starch in 200 c. c. of cold water and then boiling for a few minutes. Cool and take a few c. c. in a test tube with an equal volume of saliva. Keep warm at a temperature not exceeding 40° C. for fifteen minutes by passing the test tube through the steam from boiling water. At the end of this time test the liquid for sugar by means of Fehling's solution. The reduction of the Fehling's solution is proof of the presence of sugar produced by the action of ptyalin in the saliva upon the starch solution.

The activity of this ferment is destroyed by heat, by acids and by alkalies. This can be proved by repeating the above experiment, having previously boiled the saliva or added a drop of strong acid or alkali to it. No test for sugar will be obtained in these experiments. Another important fact to be noted is that saliva has practically no action upon uncooked starch.

6. TEST FOR SULPHOCYANATES.

Small quantities of alkali sulphocyanates exist in the saliva. To detect them proceed as follows: To a small quantity of clear saliva in a test tube add a drop of very dilute ferric chloride. A slight reddish color, due to the formation of ferric sulphocyanate, is formed, which may be removed by the addition of mercuric chloride. A comparative test can be made by adding an equal quantity of the ferric chloride solution to dilute solutions of potassium sulphocyanate.

7. TESTS FOR CHLORIDES, PHOSPHATES AND SULPHATES.

The chlorides in the saliva exist mainly as potassium and sodium chlorides, while the phosphates and sulphates are salts of both alkali and alkaline earth metals.

Before making the tests for these constituents evaporate 10 c. c. of saliva to dryness, char the organic matter, add a little distilled water and a drop or two of acetic acid and filter. Divide the filtrate into four parts.

1. For Chlorides. Test one portion for chlorides as directed under urine analysis.

2. For Phosphates. Test another portion for phosphates by adding a drop of nitric acid and an excess of ammonium molybdate and boiling. A yellow precipitate, ammonium phospho-molybdate, indicates phosphates.

3. For Sulphates. Test a third portion for sulphates by adding a drop of hydrochloric acid and some barium chloride. A white precipitate of barium sulphate indicates sulphates.

8.   TESTS FOR CALCIUM AND MAGNESIUM.

1. For Calcium.   To a fourth portion of the solution used in the above tests add ammonium oxalate. A white precipitate is calcium oxalate.   Filter.

2. For Magnesium.   To the filtrate obtained above add ammonium hydroxide to alkaline reaction. A white precipitate consists chiefly of ammonium magnesium phosphate.

9.   TEST FOR MERCURY.

Mercury is not a constituent of normal saliva, but may be present resulting from the use of drugs containing mercury.   To test for mercury proceed as follows: Collect an abundant quantity of saliva and make it acid with dilute hydrochloric acid.   Digest with gentle heat for about two hours, adding occasionally a drop of nitric acid or a fragment of potassium chlorate.   Filter and concentrate the filtrate to one-eighth its original volume.   This may be tested for mercury by removing a small quantity in a test tube, adding a small piece of bright copper wire and boiling.   The mercury will deposit on the copper.

EFFECTS OF SALIVA UPON METALS.   GALVANIC ACTION.

Galvanic action is often observed in the mouth, although it is very slight when the saliva is in a normal state.   Thus if a nail, a pin or a metallic toothpick is brought in contact with a filling in a tooth a sensation varying from a "metallic" taste in some instances to a decided shock in others is experienced.

Here the saliva acts as the exciting liquid, and when it is strongly acid or alkaline galvanic action may result sufficient to bring about the corrosion or even the destruction of metals used for various purposes in the mouth.   This is particularly true if the metals coming in contact possess wide electrical differences, as aluminum and amalgam or gold.   The subject of galvanic action in the mouth has not received enough study to lead to much definite knowledge concerning it.

Appendix.

Tables, Rules, Etc.

APPENDIX.

SECTION I.

I. Tables of Weights and Measures.

In scientific work the French or metric system of weights and measures is used almost exclusively. Since other systems are in common use, however, the relations of these to the metric system are given.

METRIC MEASURES OF WEIGHT.

1 milligram		= 0.001 gm. = 0.01543 grain.
10 milligrams	= 1 centigram	= 0.01 gm. = 0.1543 grain.
10 centigrams	= 1 decigram	= 0.1 gm. = 1.543 grains.
10 decigrams	= 1 gram	= 15.43 grains.
10 grams	= 1 dekagram	= 154.324 grains.
10 dekagrams	= 1 hectogram	= 100 gms. = 3.21 troy ounces.
10 hectograms	= 1 kilogram	= 1000 gms. = 32.1 troy ounces.

The unit of this system is the gram. It is the weight of one cubic centimeter of pure water *in vacuo* at 4° C. In scientific work only the milligram, mg., the gram, gm., and the kilogram, Kilo., are commonly used, and all other denominations are expressed in decimals. Thus, if an object weighs two grams and five decigrams, the weight would be expressed as 2.5 grams, etc.

TROY WEIGHT.

1 grain		=	0.0648 gram.
24 grains	= 1 pennyweight	=	1.55 grams.
20 pennyweights	= 1 ounce	=	31.10 grams.
12 ounces	= 1 pound	=	373.2 grams.

APOTHECARIES' WEIGHT.

1 grain		=	0.0648 gram.
20 grains	= 1 scruple	=	1.296 grams.
3 scruples	= 1 drachm	=	3.888 grams.
8 drachms	= 1 ounce	=	31.10 grams.
12 ounces	= 1 pound	=	373.2 grams.

AVOIRDUPOIS WEIGHT.

16 drams	= 1 ounce	=	28.35 grams.
16 ounces	= 1 pound	=	453.59 grams.
25 pounds	= 1 quarter	=	11.34 kilograms.
4 quarters	= 1 hundredweight	=	45.36 kilograms.
1 avoirdupois pound		=	7000 grains.

METRIC MEASURES OF VOLUME.

1000 cubic millimeters	= 1 cubic centimeter	= 0.061 cubic inch.
1000 cubic centimeters	= 1 liter	= 61.027 cubic inches.
1 cubic centimeter		= 16.23 minims.
1 liter		= 33.8 fluid ounces.
28.32 liters		= 1 cubic foot.

The units usually employed in measuring liquids in scientific work are the cubic centimeter, c. c., and the liter, L.

FLUID MEASURE.

1 mimim		=	0.0616 cubic centimeter.
60 minims	= 1 fluid drachm	=	3.696 cubic centimeters.
8 fluid drachms	= 1 fluid ounce	=	29.57 cubic centimeters.
16 fluid ounces	= 1 pint	=	473.179 cubic centimeters.
8 pints	= 1 gallon	=	3.785 liters.
	1 gallon	=	231 cubic inches.

1 millimeter	= 0.03937	inch.
10 millimeters = 1 centimeter	= 0.3937	inch.
10 centimeters = 1 decimeter	= 3.937	inches.
10 decimeters = 1 meter	= 39.37	inches.
1 inch = 25.4	millimeters.	
1 foot = 0.3048	meter.	
1 yard = 0.9144	meter.	

The units commonly employed in the metric measures of length are the millimeter, mm., the centimeter, cm. and the meter, M.

II. Rules for the Conversion of Centigrade and Fahrenheit Degrees.

Two thermometric scales are used, namely, the Centigrade and the Fahrenheit. The Centigrade is almost exclusively used in scientific work. They are mutually convertible by the following rules :

1. To Convert Centigrade into Fahrenheit; Multiply by 9, divide by 5 and add 32.

2. To Convert Fahrenheit into Centigrade: Subtract 32, multiply by 5 and divide by 9.

III. Rules for Determining and for Raising and Lowering the Carat of Gold Alloys.

1. To Determine the Carat of a Given Alloy. Multiply 24 by the weight of gold in the alloy and divide the product by the total weight of the alloy.

Example : Determine the carat of the following alloy:

Pure Gold 9 parts.
Copper   2 ''
Silver   1 ''
          ──
          12

The calculation will be 24x9÷12＝18. Hence the alloy is 18 carats fine.

In case an alloy of gold has been used in preparing the alloy, it will be necessary to first determine the quantity of gold present.

Example : Determine the carat of the following alloy :

Gold (22-carat) 9 parts.
Copper   2 ''
Silver   1 ''
              ──
              12

Since the gold is only 22 carats fine, eleven-twelfths of it is gold. Hence the alloy contains 8.25 parts of pure gold, and the statement will be 24 × 8.25÷12＝16.5. Hence the alloy is 16.5 carats fine.

2. To Raise Gold to a Higher Carat. Multiply the weight of the material taken by the difference between its carat and that of the metal to be added ; then divide this product by the difference between the carat of the metal to be added and that of the required alloy. The quotient is the total weight of the required alloy. Subtract from this the weight of the material taken, and the difference is the weight of pure or alloyed gold to be added.

Example: Raise 10 grams of 18-carat gold to 20-carat with pure gold.

$$24—18=6$$
$$24—20=4$$

$10 \times 6 \div 4 = 15$ and $15—10=5$.   Hence 5 grams of pure gold must be added to the 10 grams of 18-carat gold to raise it to 20-carat.

If, instead of adding pure gold, it is desired to use, say 22 carat, then the calculation would be:

$$22—18=4$$
$$22—20=2$$

$10 \times 4 \div 2 = 20$ and $20—10=10$.   Hence 10 grams of 22-carat gold must be used in this case.

3. *To Reduce Gold to a Required Carat.*   Multiply the weight of material used by its carat and divide this product by the required carat.   The quotient is the weight of the required alloy.   From this subtract the weight of the material used, and the difference represents the weight of inferior metal to be added.

Example: Reduce 10 grams of pure gold to 18-carat.

$10 \times 24 \div 18 = 13.3$ and $13.3—10=3.3$.   Hence 3.3 grams of inferior metal must be added.

If the gold to be reduced is not pure, but say 22 carats fine, then the calculation will be

$10 \times 22 \div 18 = 12.2$ and $12.2—10=2.2$.   Hence 2.2 grams of inferior metal must be added in this case.

IV. Standard Potassium Cyanide Solution.

To prepare and standardize a potassium cyanide solution for the volumetric estimation of copper proceed as follows: Dissolve about 60 grams of potassium cyanide in 1000 c. c. of distilled water. When in solution shake thoroughly and transfer to a dry glass stoppered bottle, which should be kept in a dark place.

Now weigh out very accurately 0.25 grams pure copper foil or wire and place in a small beaker. Add about 10 c. c. strong nitric acid and an equal quantity of water. Boil until all the copper is in solution and the brown fumes have disappeared. Transfer to a 100 c. c. flask, washing the beaker several times and adding the washings to the flask. Dilute to the mark and shake thoroughly. Draw out 50 c. c. with a pipette (equal to 0.125 of a gram of copper), transfer to a beaker and add ammonium hydroxide in slight excess. When cool, run in the potassium cyanide solution from a 50 c. c. burette with constant stirring until the blue color is partially destroyed, then add about an equal quantity of water and con- tinue the addition of potassium cyanide until the blue color is destroyed and the color has changed to a faint pink. Repeat the operation on the other half of the copper solution and take the average of the number of cubic centimeters employed.

Example : If it took 18.2 c. c. potassium cyanide solution for 0.125 of a gram of copper (contained in 50 c. c. of copper solution), then $0.125 \div 18.2 = 0.00686$. Hence 1 c. c. of the cyanide solution is equivalent to 0.00686 of a gram of copper.

V. Fehling's Solution.

Fehling's solution, used in the detection and esti- mation of sugar, is a copper sulphate solution containing an alkali and a tartrate. Since Fehling's solution does not keep well it is best to retain the copper sulphate solution and the alkaline tartrate solution in separate bottles. When Fehling's solution is needed it can be made by mixing equal volumes of these two solutions.

1. Copper Sulphate Solution. Dissolve 69.28 grams of pure crystallized copper sulphate in 500 c. c. of distilled water and then dilute to make 1000 c. c. of solution.

2. Alkaline Tartrate Solution. Dissolve 100 grams of stick sodium hydroxide in 500 c. c. of distilled water. When the alkali is dissolved add slowly with heat 350 grams of pure Rochelle salt and stir until dissolved. Allow the solution to stand for twenty-four hours and then filter it through glass wool. Finally, add enough water to bring the volume up to 1000 c. c.

VI. Sodium Hypobromite Solution.

The sodium hypobromite solution used in determining the quantity of urea in urine is prepared as indicated below. Since it does not keep well, it is best to prepare a fresh solution when required.

Dissolve 100 grams of stick sodium hydroxide in 250 c. c. of water. When cool, place the vessel containing this solution in cold water and add slowly with constant stirring 25 c. c. of bromine. This should be done in a good draft, as the fumes of bromine are very irritating. Finally, dilute this solution with an equal volume of water before using.

SECTION II.

I. Forms of Students' Report Sheets.

It is the custom of the author to require students to make reports upon the work done in the laboratory, and printed report sheets are furnished for this purpose. The forms for the various exercises are shown below :

REPORT ON THE ANALYSIS OF UNKNOWN SUBSTANCES.

Name.......... Section...... Date...........
Solution, metal or alloy......................
Metals found.

....

....

Remarks :

....

REPORT ON THE REFINING OF GOLD, SILVER AND MERCURY.

Name........... Section..... Date. ....
Weight of scrap gold taken ....
Character of scrap...........................
Method employed.

....

Weight of pure gold obtained...............
Weight of scrap silver, or of amalgam, taken.

....

Methods employed...........................

....

Weight of each metal obtained.
Remarks :

REPORT ON AMALGAM-ALLOYS AND AMALGAMS.

Name.........Section.....Date............

Formula of amalgam-alloy. ....

Weight of each metal used...................

Weight of ingot obtained.....................

Loss in melting...........................

1. Weight *in air* of each portion of ingot.

   No. 1........ No. 2........ No. 3........

   Weight *in water* of each portion of ingot.

   No. 1........ No. 2........ No. 3........

   Specific gravity of each portion of ingot.

   No. 1........ No. 2........ No. 3........

   Average specific gravity of whole ingot........

   Theoretical specific gravity of whole ingot.....

   How cut—filings or shavings...............

   Filings.   Fine.....Medium.....Coarse.....

Tests on Freshly Cut Alloy (or Annealed or Oxidized).

2. Discoloration.   Light....Medium....Dark ...

3. Weight of filings taken.....................

   Weight of mercury taken...................

   Setting property.   Quick...Medium...Slow...

   Character of margins when packed (under microscope)..........................

   Character of margins after several days (under microscope)...........................

Micrometrical measurements.

Time...... Expansion...... Contraction......

....

....

....

4. Crushing-strength. .... (Lbs.)

   Flow... (Per cent.) Time... Pressure.. (Lbs.)

5. Weight of mercury expressed................

   Percentage composition of amalgam..........

   Weight of tin removed by mercury...........

   Percentage of tin in expressed mercury.......

   Percentage of the total weight of tin in the

alloy removed....................................

   Remarks:

....

....

....

....

....

Copper Amalgam.

Remarks:

APPENDIX.

REPORT ON ASSAY OF AMALGAM-ALLOY.

Name.......... Section...... Date. ....

Percentage composition of alloy as determined by assay.
Silver..... Zinc.......
Tin....... Gold.......
Copper..... Total....

Percentage composition of alloy calculated from weight of each metal taken in preparing the alloy.
Silver..... Zinc ......
Tin ..... Gold.......
Copper. .... Total....

Remarks :

....

....

REPORT ON PREPARATION OF SOLDERS.

Name.......... Section..... Date.............
Solder or solders prepared.....................
Quantities prepared.........................
Remarks :

....

....

REPORT ON MISCELLANEOUS ALLOYS.

Name.......... Section..... Date.............
Fusible alloys prepared.......................
Melting points of alloys as determined by experiment.....................................
Remarks:

....

....

REPORT ON DENTAL CEMENTS.

Name............Section.....Date............

Tests on Oxyphosphate Cement.

1. Results of *General Tests.*

    1 ....
    2 ....
    3 ....
    4 ....
    5 ....
    6 ....

2. Character of margins when packed (under microscope)................................

    Character of margins after several days (under microscope)................................

    Micrometrical measurements.

    Time ..... Expansion...... Contraction......
    ....
    ....

3. Crushing-strength...................(Lbs.)

4. Results of *Porosity Tests.*

    ....

5. Results of *Arsenic Test.*

    ....

    Remarks:

....

Oxychloride Cement.

    Remarks:

....

Oxysulphate Cement.

    Remarks:

....

REPORT ON SPECIAL PROBLEMS.

Name............Section.....Date...........

Analysis of Teeth.

Quantitative tests.
1. Weight of sample taken...................
    Weight after dehydration..................
    Loss due to water. ....
    Percentage of water. ....
2. Weight of residue after calcining...........
    Loss due to organic matter. ....
    Percentage of organic matter. ....
    Percentage of inorganic matter. ....
Qualitative tests.
1. Results of *Tests for Organic Matter*........
2. Results of *Tests for Phosphates.* ....
3. Results of *Tests for Calcium and Magnesium Phosphates.* ....
4. Results of *Test for Calcium*..................
Remarks:
....

Analysis of Urine.

1. Physical characteristics...................
2. Specific gravity. ....
3. Results of *Tests for Albumin* ....
4. Results of *Test for Mucin.*.................
5. Results of *Tests for Sugar*.................
6. Results of *Tests for Abnormal Coloring Matter.*
    ....
7. Results of *Tests for Phosphates and Chlorides.*
    ....

8.   Weight of uric acid present in 1500 c. c......

9.   Weight of urea present in 1500 c. c.........

Remarks:

....

Analysis of Saliva.

1.   Physical characteristics....................

2.   Specific gravity. ....

3.   Results of *Tests for Albumin*...............

4.   Results of *Test for Mucin*.. ....

5.   Results of *Tests for Ptyalin*.............. ....

6.   Results of *Test for Sulphocyanates* ....

7.   Results of *Tests for Chlorides, Phosphates and Sulphates*......- ....

8.   Results of *Tests for Calcium and Magnesium.*

....

9.   Results of *Test for Mercury* ....

Remarks :

....••

II. COMPOSITION OF SOME WELL-KNOWN DENTAL CEMENT POWDERS (BY THE AUTHOR).

Number.	Insoluble matter.	Zinc Oxide.	Tin Oxide.	Bismuth Oxide.	Arsenious Oxide.
1	Trace	100.0	…	…	0.04
2	0.5	99.5	…	…	0.05
3	0.5	90.0	9.5	…	0.04
4	6.0	94.0	…	…	0.09
5	0.8	99.2	…	…	0.04
6	4.4	95.6	…	…	Trace
7	1.2	98.8	…	…	Trace
8	Trace	100.0	…	…	0.04
9	1.2	88.6	…	10.2	Trace
10	Trace	100.0	…	…	Trace
11	9.6	90.4	…	…	0.09
12	Trace	100.0	…	…	Trace

III. COMPOSITION OF SOME WELL-KNOWN AMALGAM-ALLOYS.*

NAME.	Tin	Silver	Gold	Platinum	Copper	Zinc	Cadmium	Antimony	Palladium
Arrington's (S. S. White's)	57.5	42.5	0.5	0.15		0.5			
Blackwood's G. and P. Alloy	56.85	42.		0.5					
Best (Spencer & Crocker's) Old	61.5	34.5	6.		3.5				
Chicago Refining Co.'s (Old)	56.	37.	4.	2.	10.				
" " (New)	58.97	37.55		0.1		.			
Chase's Coppered Amalgam	50.	50.							
" Plastic Tin Amalgam	50.	50.						6	
" Alcoholic Tight Amalgam	40.	50.						7	
" Stannous Gold	40.	40.	20.					10	
" Incisor Tooth Amalgam	40.	50.						10	
Caulk's White Alloy	55.	43.65	0.16	0.25	1.35	2.44			
" Par-Excellence	61.75	27.25	0.15		10.6	1.45			
Crown Gold Alloy	52.85	47.	0.05			0.4			
Dawson's White Alloy	49.27	48.21	0.65			1.05			
" Superior Amalgam	63.65	31.86	0.2	0.15	2.35				Trace.
Dibble's White Amalgam	49.65	39.75							
Fry's Amalgam	53.8	44.35	4.	0.9					
Fletcher's Gold Alloy (Old)	66.	40.	3.35	1.3					
" Platinum and Gold Alloy	60.35	41.35							
Flagg's Submarine	35.	60.	6.		1.65				
" Facing	35.	57.	6.		6.	3.			
" Contour Alloy	37.	58.	1.5						
Globe (S. S. White's)	53.36	44.74		0.4					
Grimes's Front Tooth (Old)	44.	10.					46		
Hood's Amalgam (Old)	60.25	37.	2.75			1.2			
Hood & Reynold's Gold and Platinum Alloy	60.4	44.3	3.8	0.3		1.8			
" " Sans Tache Alloy	50.	47.9			0.3				
Holmes's Star No. 1 (Old)	59.	40.	1.	0.5					
" " 2 (New)	58.5	39.5	2.	1.					
Hays's Pure White (Old)	51.5	43.5				4.			
Hardman's Amalgam	41.57	50.12	0.28		5.31				
" White Alloy	50.56	44.57			4.59				

* "American System of Dentistry,"

COMPOSITION OF SOME WELL-KNOWN AMALGAM-ALLOYS.*—Continued.

Name.	Tin	Silver	Gold	Platinum	Copper	Zinc	Cadmium	Antimony	Palladium
High Grade Alloy (7½ per cent Gold)	41.5	49.	7.5			2.			
Harris's (Prof. J. H.) Amalgam	48.1	40.			4.9	7.			
Johnson & Lund's Extra (Old)	60.	38.							
" " " (New)	61.15	36.75	1.5	0.5		0.6	1.45		
" " Virgin White Alloy	61.65	37.75	0.15	0.5			0.9		
" " Atlas Analgam	61.9	36.85							
" " Extra Tough Alloy	51.25	47.	0.3	0.35	0.25				
Justi's Superior Gold and Platinum Alloy	59.1	35.2	0.32	0.2	3.5	1.8			
King's Occidental	54.75	42.75		0.08		2.5			
Lawrence's (Old)	47.	47.	1.		5.				
" (New)	50.43	44.06			5.51				
Moffit's (Old)	62.	36.	2.						
(Moffit's) The Dentist's Amalgam	59.5	37.9				2.6			
Oliver's Amalgam (Old)	50.8	46.1	1.7	1.4					
" White Amalgam (New)	55.25	44.74							
Peirce's (Old)	40.	55.	4.			1.			
Prof. Essig's (Old)	55.	45.	2.5	2.5					
Parson's Eureka Silver Alloy	40.	55.			3.	2.			
Sterling Amalgam (Old)	62.	31.2	1.		6.				
" D. & L. (New)	62.37	33.2		0.14	4.93				
Standard Amalgam (Davis & Co.)	55.4	44.6							
Dental Alloy (Eckfeldt)	40.6	52.	4.4	0.08	3.				
Shattuck's Standard Gold Alloy	51.74	46.98	1.2						
Sibley's Gold and Platinum Alloy	54.65	43.15	0.2	2.					
Temporary Alloy	88.	10.				2.			
Townsend's (Old)	58.	42.							
" (Improved)	54.5	44.5	1.						
Walker's (Old)	69.	30.5							
" Excelsior Gold and Platinum Alloy	51.5	42.	0.3	0.5	6.				
Welch's Gold and Platinum Alloy (Old)	54.	44.	1.3	0.2					
" " " (New)	51.9	46.	1.7	0.7					
" Amalgam	51.52	48.48		0.4					

* "American System of Dentistry."

INDEX.

The Great Forward Movement in Dentistry

grows apace

This is apparent alike in the requirements as to the qualifications and work of the students entering our colleges, the graduates leaving them, the professional instructors and private practitioners. ALL feel the mighty impulse derived from strengthened vision and the high altitudes from which they view the masterly possibilities of the dental profession in this dawn of the second century of our national life.

We believe it may be truthfully said that we, as manufacturers and dealers, have kept abreast of and much of the time anticipated the demands of the profession, in furnishing high grade tools and materials.

New evidence of alertness appeared recently when that scientific investigator and indefatigable worker, Dr. G. V. Black, M. D., D. D. S., Sc. D., Dean of the Dental Department of the Northwestern University, so well known and greatly respected wherever dentistry is practiced, conceived and successfully conducted a course in "Amalgam Making" in Chicago.

Our chemist and metallurgist was promptly on hand, took the course, received his certificate, and we will, a little later, place upon the market an amalgam made in comformity with Dr. Black's ideas and meeting fully his requirements.

You will find at our Chicago house a complete stock of up-to-date dental supplies. Our representatives will be found to be experienced and courteous, with exceptional facilities for supplying your needs from the least to the largest item of dental requirements.

We serve you courteously, conscientiously, carefully and promptly. Try us.

GIDEON SIBLEY

N. W. Cor. State and Quincy Sts.

PHILADELPHIA, PA.   Chicago, Ill.

www.ingramcontent.com/pod-product-compliance
Lightning Source LLC
Chambersburg PA
CBHW021519210326
41599CB00012B/1311